Estimating Ore Grade Using Evolutionary Machine Learning Models

Mohammad Ehteram · Zohreh Sheikh Khozani ·
Saeed Soltani-Mohammadi · Maliheh Abbaszadeh

Estimating Ore Grade Using Evolutionary Machine Learning Models

Springer

Mohammad Ehteram
Department of Water Engineering
and Hydraulic Structures
Faculty of Civil Engineering
Semnan University
Semnan, Iran

Zohreh Sheikh Khozani
Faculty of Civil Engineering
Institute of Structural Mechanics
Bauhaus University Weimar
Weimar, Germany

Saeed Soltani-Mohammadi
Department of Mining Engineering
University of Kashan
Kashan, Iran

Maliheh Abbaszadeh
Department of Mining Engineering
University of Kashan
Kashan, Iran

ISBN 978-981-19-8105-0 ISBN 978-981-19-8106-7 (eBook)
https://doi.org/10.1007/978-981-19-8106-7

This Springer imprint is published by the registered company Springer Nature Singapore Pte Ltd.
The registered company address is: 152 Beach Road, #21-01/04 Gateway East, Singapore 189721, Singapore

Preface

In mining engineering and geoinformatics, grade estimation is one of the most important topics. Developing new models for grade estimation is one of the most important issues of mining engineering. Estimating ore grade is necessary for planning and analyzing mine finances. Grade estimation is an important step in assessing the quality of a mineral deposit. The mineral industry requires a reliable tool for estimating ore grade. Poor ore grade estimation negatively impacts mine valuation and pit optimization. In this book, new and developed methods for estimating grade are presented. The current book presents developed artificial neural network models for grade estimation in various case studies. This book introduces models and methods that can reduce mining and drilling costs.

Additionally, the models used in this book can speed up the modeling process. Engineers, students, modelers, industrial managers, and researchers can use the current book's models to estimate various variables. This book contains models that can be used by researchers in other fields as well as mining engineers.

Chapter 1 discusses the importance of ore grade estimation. Chapter 2 reviews the different methods for estimating the grade. Chapter 3 introduces different neural network models. Training and optimization algorithms for estimating grades are presented in Chap. 4. Chapters 5–8 use evolutionary artificial neural networks to estimate the ore grade. The conclusion is presented in Chap. 9.

Semnan, Iran
September 2022

Mohammad Ehteram

Contents

List of Figures

List of Tables

Chapter 1
The Necessity of Grade Estimation

Abstract Estimating ore grade is one of the most important aspects of mining engineering. This chapter aims to explain the importance of estimating grades. Modelers need robust models for estimating ore grade since it is a nonlinear and complex process. We investigate the potential of different models for estimating ore grade. We explain the advantages and disadvantages of models. The purpose of this chapter is to assist modelers in choosing the best model for estimating ore grade. The modelers need to understand the advantages and disadvantages of the different models before choosing a robust model for estimating ore grade. This chapter introduces artificial intelligence models as reliable tools for estimating ore grades.

Keywords Ore grade estimation · Artificial intelligence models · Kriging models · Mining engineering

1.1 Introduction

The purpose of this section is to explain the importance of estimating ore grades. The next step is to describe classical models for estimating ore grades. Next, we discuss the advantages of new ore grade estimation models. The last section presents general remarks about various methods of calculating ore grades.

1.1.1 The Importance of Ore Grade Estimation

As Industrial projects expand worldwide, the exploration and use of mineral deposits will increase. When investing in new projects or opening new mines, it is essential to evaluate mineral deposits. Geologists and mining engineers continuously explore ways to understand and simulate earth processes (Abbaszadeh et al., 2021). Estimating ore grades is an important step in the qualitative evaluation of mines (Saljoughi & Hezarkhani, 2018).

Mineral concentrations (grade) depend on the location, the geological composition, and the structural features of the ore deposit (Jafrasteh et al., 2018). Planning,

M. Ehteram et al., *Estimating Ore Grade Using Evolutionary Machine Learning Models*, https://doi.org/10.1007/978-981-19-8106-7_1

production schedules, and feasibility of mining projects depend on the accuracy of grade estimation. (Li et al., 2013). For successful and profitable mining, an accurate grade estimation is necessary (Tahmasebi and Hezarkhani, 2012). Mine development relies mainly on accurately predicting ore grade. Mining companies always strive to develop reliable models for estimating grades. Mining operations are adversely affected by an inaccurate ore grade model.

1.2 Conventional Ore Grade Estimation Models

The estimation of ore grade is influenced by many factors, such as sampling variability and the natural characteristics of an ore deposit (Jafrasteh et al., 2018). Geometrical, distance-weighting, geostatistical, and machine learning methods have been developed to estimate ore grade.

The geometrical, distance weighting, and geostatistical approaches are categorized as conventional methods, and artificial intelligence methods are considered modern models. Geometrical approach techniques include polygonal, triangular, and cross-sectional estimation methods. These methods are usually applied when sufficient data are unavailable at an exploratory stage (Afeni et al., 2021). Inverse distance weighting (IDW) is one of the most important techniques (Abuntori et al., 2021; Sinclair & Blackwell, 2002). Nowadays, geostatistical methods are successfully used for grade estimation. Geostatistical methods enable modelers to obtain more precise information about grade variability (Kokesz, 2006).

Various geostatistical methods are applied for grade estimation, such as ordinary kriging (Afzal, 2018; Daya, 2015), log-kriging (DENG et al., 2007; Thurston & Armstrong, 1987), disjunctive kriging (Yu et al., 2004), multiple indicator kriging (Kasmaee & Torab, 2014; Lipton et al., 1998; Soltani-Mohammadi & Erhan Tercan, 2012), fuzzy kriging (Bardossy et al., 1990; Soltani-Mohammadi, 2016), localized uniform conditioning (Abzalov, 2006), etc. Among geostatistical methods, ordinary kriging is a popular estimation method in the mining industry. The Kriging method minimizes estimation variance by calculating optimal weights for samples. (Journel & Huijbregts, 1978). Geostatistical interpolation methods are beyond the scope of this book, and readers are referred to standard textbooks for more information. (Journel & Huijbregts, 1978; Rossi & Deutsch, 2014; Webster & Oliver, 2001). Geostatistical methods use assumptions about the spatial correlation of ore grades in a mineral deposit. Because of these assumptions, the techniques become complicated (Abuntori et al., 2021). Geostatistical methods may not provide accurate estimates when grade distribution and spatial pattern variability are complex.

Geostatistical methods require enough data to develop variogram models that accurately describe mineralization (Annels, 2012). Because of these limitations and complexities, researchers explore alternative methods to overcome obstacles (Abuntori et al., 2021). This section indicated that the conventional methods have drawbacks. It is necessary to develop a new method of estimating ore grade to overcome the limitations of conventional methods, such as geostatistical analyses.

1.3 New Models for Estimating Ore Grade

Artificial intelligence (AI) has gained incredible popularity in the last decade. Leading companies have invested in this area to address the challenges in various industries (Ehteram et al., 2021a). In addition, governments have increased funding for AI research centers (Ehteram et al., 2021b). Machine learning models are the most important sub-models of artificial intelligence models (Mirjalili et al., 2020). In computer science, machine learning models deal with the learning aspect of computers (Mirjalili et al., 2020). There are three kinds of learning methods: supervised, unsupervised, and reinforcement algorithms (Mirjalili et al., 2020). Unsupervised algorithms include the following methods (Mirjalili et al., 2020; Seifi et al., 2020):

(1) K-means clustering: Data points are divided into multiple classifications based on similar characteristics.
(2) Self-organizing Neural Networks: The neurons of this neural network are arranged to minimize an error function (Khozani et al., 2022; Seifi et al., 2022).

A supervised learning method can check whether a particular action is correct or incorrect. These are some popular supervised machine learning algorithms:

(1) Support vector machine: A support vector machine finds a hyperplane in N-dimensional space that uniquely classifies the data points (Essam et al., 2022; Banadkooki et al., 2019; Adnan et al., 2022; Mohamadi et al., 2020).
(2) Random forest: This supervised learning method uses multiple decision trees (made up of features) and integrates them using a merit factor.
(3) Artificial neural networks: The neurons in this learning method are arranged in layers and are connected using weight connections (Panahi et al., 2021).

Reinforcement involves rewarding correct actions and punishing incorrect actions (Mirjalili et al., 2020).

(1) Q-learning: The learning method maximizes Q-value by using the Bellman equation.
(2) Deep Q Network: Unlike Q learning, this method is more generalized.

Supervised learning machine models are widely used for grade estimation (Abbaszadeh et al., 2021). The advantages of supervised machine learning models include ease of implementation, fast convergence, high accuracy, and flexibility. It is complex and nonlinear to develop a model for estimating ore grade (Abbaszadeh et al., 2021). Thus, machine learning algorithms can help modelers estimate ore grades accurately. For mining engineers, accurate grade estimation is an important topic (Soltani-Mohammadi et al., 2022). It is difficult for modelers to estimate ore grade precisely due to several uncertainties. Machine learning models can handle complex, nonlinear, and uncertain data. Due to these features, machine learning models can accurately estimate ore grade. The next chapter reviewed machine learning models for estimating ore grades (Soltani-Mohammadi et al., 2022).

1.4 General Remarks

Mineral deposit evaluation is a critical task in mining engineering. Estimating ore grade is one of the most important stages of mineral resource evaluation. Using robust models to estimate ore grade is necessary because geological and chemical characteristics affect it. The conventional models can model ore grade, but they have drawbacks. They may not handle complex boundary conditions and complex case studies. Such models may also be time-consuming. Machine learning algorithms have a high ability to estimate ore grade due to their advantages. These models employ artificial intelligence to determine the relationship between ore grade and input data.

References

Adnan, R. M., Kisi, O., Mostafa, R. R., Ahmed, A. N., & El-Shafie, A. (2022). The potential of a novel support vector machine trained with modified mayfly optimization algorithm for streamflow prediction. *Hydrological Sciences Journal*. https://doi.org/10.1080/02626667.2021.2012182.

Abbaszadeh, M., Ehteram, M., Ahmed, A. N., Singh, V. P., & Elshafie, A. (2021). The copper grade estimation of porphyry deposits using machine learning algorithms and henry gas solubility optimization. *Earth Science Informatics*. https://doi.org/10.1007/s12145-021-00667-6

Abuntori, C. A., Al-Hassan, S., & Mireku-Gyimah, D. (2021). Assessment of ore grade estimation methods for structurally controlled vein deposits - a review. *Ghana Mining Journal, 21*(1), 31–44. https://doi.org/10.4314/gm.v21i1.4

Abzalov, M. Z. (2006). Localised uniform conditioning (LUC): A new approach for direct modelling of small blocks. *Mathematical Geology, 38*(4), 393–411.

Afeni, T. B., Akeju, V. O., & Aladejare, A. E. (2021). A comparative study of geometric and geostatistical methods for qualitative reserve estimation of limestone deposit. *Geoscience Frontiers, 12*(1), 243–253. https://doi.org/10.1016/j.gsf.2020.02.019

Afzal, P. (2018). Comparing ordinary kriging and advanced inverse distance squared methods based on estimating coal deposits; case study: East-Parvadeh deposit, central Iran. *Journal of Mining and Environment*, *9*(3), 753–760. https://doi.org/10.22044/jme.2018.6897.1522

Annels, A. E. (2012). *Mineral deposit evaluation: A practical approach.* Springer Netherlands. Retrieved from https://books.google.com/books?id=uevqCAAAQBAJ

Banadkooki, F. B., Ehteram, M., Ahmed, A. N., Fai, C. M., Afan, H. A., Ridwam, W. M., & El-Shafie, A. (2019). Precipitation forecasting using multilayer neural network and support vector machine optimization based on flow regime algorithm taking into account uncertainties of soft computing models. *Sustainability, 11*(23), 6681.

Bardossy, A., Bogardi, I., & Kelly, W. E. (1990). Kriging with imprecise (fuzzy) variograms. II: Application. *Mathematical Geology*, *22*(1), 81–94. https://doi.org/10.1007/BF00890298

Daya, A. (2015). Ordinary kriging for the estimation of vein type copper deposit: A case study of the Chelkureh, Iran. *Journal of Mining and Metallurgy A: Mining, 51*, 1–14. https://doi.org/10.5937/JMMA1501001D

Deng, M. G., Li, W. C., Bo, LI., Li, L. H., Jiang, S. D., Xiong, G. X., Zhang, X. S., & Yu, H. J. (2007). Application of log kriging on estimated reserves of the 10-9 ore body of lutangba in the gejiu tin deposits. *Journal of China University of Mining and Technology*, *17*(2), 286–289.

Ehteram, M., Ahmed, A. N., Kumar, P., Sherif, M., & El-Shafie, A. (2021a). Predicting freshwater production and energy consumption in a seawater greenhouse based on ensemble frameworks using optimized multi-layer perceptron. *Energy Reports, 7*, 6308–6326. https://doi.org/10.1016/j.egyr.2021.09.079

Ehteram, M., Sammen, S. S., Panahi, F., & Sidek, L. M. (2021b). A hybrid novel SVM model for predicting CO2 emissions using multiobjective seagull optimization. *Environmental Science and Pollution Research, 28*(46), 66171–66192. https://doi.org/10.1007/s11356-021-15223-4

Essam, Y., Huang, Y. F., Ng, J. L., Birima, A. H., Ahmed, A. N., & El-Shafie, A. (2022). Predicting streamflow in Peninsular Malaysia using support vector machine and deep learning algorithms. *Scientific Reports*. https://doi.org/10.1038/s41598-022-07693-4.

Jafrasteh, B., Fathianpour, N., & Suárez, A. (2018). Comparison of machine learning methods for copper ore grade estimation. *Computational Geosciences, 22*(5), 1371–1388.

Journel, A. G., & Huijbregts, C. J. (1978). *Mining geostatistics*. Academic Press.

Kasmaee, S., & Torab, F. M. (2014). Risk reduction in Sechahun iron ore deposit by geological boundary modification using multiple indicator kriging. *Journal of Central South University, 21*(5), 2011–2017. https://doi.org/10.1007/s11771-014-2150-x

Khozani, Z. S., Banadkooki, F. B., Ehteram, M., Ahmed, A. N., & El-Shafie, A. (2022). Combining autoregressive integrated moving average with Long Short-Term Memory neural network and optimisation algorithms for predicting ground water level. *Journal of Cleaner Production, 348*, 131224.

Kokesz, Z. (2006). Application of linear geostatistics to evaluation of Polish mineral deposits. *Gospodarka Surowcami Mineralnymi – Mineral Resources Management, 22*(2), 53–65. Retrieved from https://gsm.min-pan.krakow.pl/Application-of-linear-geostatistics-to-evaluation-of-Polish-mineral-deposits,96645,0,2.html

Li, X-li., Li, L-hong, Zhang, B-lin, Guo, Q-jin. (2013). Hybrid self-adaptive learning based particle swarm optimization and support vector regression model for grade estimation. *Neurocomputing, 118*, 179–190. https://doi.org/10.1016/j.neucom.2013.03.002

Lipton, I., Gaze, R., Horton, J., & Khosrowshahi, S. (1998). Practical application of multiple indicator kriging and conditional simulation to recoverable resource estimation for the Halley's Lateritic Nickel Deposit. In *Beyond Ordinary Kriging: Non-Linear Geostatistical Methods in Practice, Proceedings of a 1 Day Symposium* (pp. 88–105), Perth.

Mirjalili, S., Faris, H., & Aljarah, I. (2020). *Introduction to evolutionary machine learning techniques* (pp. 1–7). Springer, Singapore. https://doi.org/10.1007/978-981-32-9990-0_1

Mohamadi, S., Sammen, S. S., Panahi, F., Ehteram, M., Kisi, O., Mosavi, A., Ahmed, A. N., El-Shafie, A., & Al-Ansari, N. (2020). Zoning map for drought prediction using integrated machine learning models with a nomadic people optimization algorithm. *Natural Hazards*. https://doi.org/10.1007/s11069-020-04180-9.

Panahi, F., Ahmed, A. N., Singh, V. P., Ehtearm, M., elshafie, A., & Torabi Haghighi, A. (2021). Predicting freshwater production in seawater greenhouses using hybrid artificial neural network models. *Journal of Cleaner Production*. https://doi.org/10.1016/j.jclepro.2021.129721.

Rossi, M., & Deutsch, C. (2014). *Mineral resource estimation*. https://doi.org/10.1007/978-1-4020-5717-5

Saljoughi, B. S., & Hezarkhani, A. (2018). A comparative analysis of artificial neural network (ANN), wavelet neural network (WNN), and support vector machine (SVM) data-driven models to mineral potential mapping for copper mineralizations in the Shahr-e-Babak region, Kerman, Iran. *Applied Geomatics, 10*(3), 229–256. https://doi.org/10.1007/s12518-018-0229-z

Seifi, A., Ehteram, M., & Soroush, F. (2020). Uncertainties of instantaneous influent flow predictions by intelligence models hybridized with multi-objective shark smell optimization algorithm. *Journal of Hydrology, 587*, 124977.

Seifi, A., Ehteram, M., Soroush, F., & Haghighi, A. T. (2022). Multi-model ensemble prediction of pan evaporation based on the Copula Bayesian model averaging approach. *Engineering Applications of Artificial Intelligence, 114*, 105124.

Sinclair, A. J., & Blackwell, G. H. (2002). *Applied mineral inventory estimation*, p. 400. ISBN 0-521-79103-0. Cambridge, UK: Cambridge University Press. https://doi.org/10.1017/CBO9780511545993

Soltani-Mohammadi, S., & Erhan Tercan, A. (2012). Constrained multiple indicator kriging using sequential quadratic programming. *Computers and Geosciences, 48*. https://doi.org/10.1016/j.cageo.2012.01.003

Soltani-Mohammadi, S. (2016). FuzzyKrig: A comprehensive matlab toolbox for geostatistical estimation of imprecise information. *Earth Science Informatics, 9*(2). https://doi.org/10.1007/s12145-015-0240-4

Soltani-Mohammadi, S., Hoseinian, F. S., Abbaszadeh, M., & Khodadadzadeh, M. (2022). Grade estimation using a hybrid method of back-propagation artificial neural network and particle swarm optimization with integrated samples coordinate and local variability. *Computers and Geosciences, 159*, 104981. https://doi.org/10.1016/j.cageo.2021.104981

Tahmasebi, P., & Hezarkhani, A. (2012). A hybrid neural networks-fuzzy logic-genetic algorithm for grade estimation. *Computers & Geosciences, 42*, 18–27. https://doi.org/10.1016/j.cageo.2012.02.004

Thurston, M., & Armstrong, M. (1987). The application of log semi-variograms to the kriging of moderately skew raw data. In *Twentieth International Symposium on the APCOM* (Vol. 3).

Webster, R., & Oliver, M. A. (2001). Geostatistics for environmental scientists. In *Statistics in practice*. Wiley.

Yu, C., Yu, X., & Hou, J. (2004). The theory of disjunctive kriging and its application in grade estimate. In *IGARSS 2004. 2004 IEEE International Geoscience and Remote Sensing Symposium*, (Vol. 6, pp. 4176–4179). https://doi.org/10.1109/IGARSS.2004.1370055

Chapter 2
A Review of Modeling Approaches

Abstract The aim of this study is to review the application of different models for estimating ore grades. We discuss the advantages and disadvantages of different models. The chapter also discusses the benefits of different soft computing models. This chapter aims to assess the potential of artificial neural networks for estimating ore grades. In addition, this chapter examines the research gaps for estimating ore grade in previous studies. Additionally, this chapter suggests solutions to improve the accuracy of models for estimating ore grades. This chapter examines the drawbacks of different models. The chapter indicated that ore grade could be accurately estimated using soft computing models.

Keywords Artificial intelligence models · Artificial neural network models · Estimating ore grade · Numerical models

2.1 Introduction

This section aims to review the application of machine learning models (MLMs) for estimating ore reserves. Next, the advantages and disadvantages of different models are highlighted. Finally, general comments are provided on estimating ore grade.

2.1.1 A Review of Studies of Applying MLMs for Estimating Ore Grade

Wu and Zhou (1993) used a multi-layer feedforward neural network to estimate copper grade. The network was built using the longitude and latitude of samples. Consequently, the network architecture consists of two nodes in the input layer. Neural networks produced more accurate results than conventional models. Additionally, they noted that this approach did not require assumptions about ore grade distribution. Koike and Matsuda (2003) predicted four impurity elements of limestone using a feedforward neural network. They defined longitude, latitude, the elevation of samples, rock types (such as crystalline and conglomeratic limestone), and

M. Ehteram et al., *Estimating Ore Grade Using Evolutionary Machine Learning Models*, https://doi.org/10.1007/978-981-19-8106-7_2

fossil types (related to the geologic age of the limestone) as input neurons. They reported that supplementary information (such as lithology and biostratigraphy) could improve spatial estimation accuracy.

Samanta et al. (2002) used a genetic algorithm to address the problem of random selection in sparse data. They divided the data set into training, validation, and testing subsets. The gold grade estimation model was developed using a multi-layer feedforward neural network and early/quick stop training. Samanta et al. (2005a) designed an artificial neural network for Al_2O_3 and Sio_2 grade estimations in a bauxite deposit. Samanta et al. (2005b) evaluated the performance of the Adaboost ensemble algorithm for estimating gold grade and compared its results with a single neural network. Chatterjee et al. (2007) coupled general regression neural network (GRNN) with the kriging model for estimating ore grade. Spatial coordinates were used as inputs to develop the models. The hybrid GRNN-kriging model performed significantly better than the ordinary kriging model. Chatterjee et al. (2008) used artificial neural networks (ANNs) to estimate the ore grades. The genetic algorithm was used to select the model parameters. This model was compared with an ordinary kriging model. Based on the results, the ANN outperformed the kriging model. Chatterjee et al. (2010) used a genetic algorithm (GA) and k-means clustering for ore grade estimation. The developed method was successful in predicting zinc grades.

Dutta et al. (2010) used ANN and support vector machine models to estimate ore grade. They concluded that the SVM and ANN models were superior to the kriging method. Tahmasebi and Hezarkhani (2012) used Coactive Neuro-Fuzzy Inference System (CANFIS) to estimate ore grade. The network parameters were optimized using GAsCANFIS-GA was more accurate and faster than other time-consuming methods. Majdifar and Kamali (2014) used a neuro-fuzzy system to estimate ore grades. The neuro-fuzzy system could predict the ore grade accurately. By combining artificial neural networks and the inverse distance weighted method, Karami and Afzal (2015) estimated ore grades. The correlation between raw data and the outputs demonstrated that the combined method could be successfully applied to estimate ore grade. Bazdar et al. (2015) combined artificial neural network models (ANN) with an invasive weed optimization (IWO) algorithm for estimating ore grades. ANN-IWO had a high capacity for estimating ore grade. Patel et al. (2017) developed a vision-based online classification technique for iron ores. A SVM model was developed to classify iron ores. An SVM classification method outperformed other classification methods. Jafrasteh and Fathianpour (2017a) combined the artificial bee colony algorithm with backpropagation to train ANN models and estimate ore grade. The results showed that the improved ANN models accurately estimated ore grade. Jafrasteh and Fathianpour (2017b) used a fuzzy-artificial bee colony algorithm to estimate ore grade. Results showed that the fuzzy-artificial bee colony algorithm predicted ore grade accurately. For calculating a deposit's grade distribution, Kaplan and Topal (2020) integrated multi-layer feedforward neural networks (NN) and k-nearest neighbor (kNN) models. The new hybrid method was a powerful tool for estimating ore grades. Abuntori et al. (2021) used five Extreme Learning Machine (ELM) variants to estimate ore grade. The ELM (ELM-Sigmoid) gave the best results.

Abbaszadeh et al. (2021) estimated the copper grade using an adaptive neuro-fuzzy inference system (ANFIS) and multi-layer perceptron (MLP). For setting the parameters of the MLP and ANFIS models, the Henry gas solubility optimization (HGSO), weed algorithm (WA), and moth flame optimization (MFO) methods were used. The optimization algorithms improved the accuracy of ANFIS and MLP models. Soltani-Mohammadi et al. (2022) combined backpropagation with particle swarm optimization (PSO) for ore grade estimation. They reported that BP-PSO had a high ability to calculate ore grade and train ANN models. Even though previous studies showed the high performance of machine learning algorithms, they had shortcomings and limitations. These shortcomings and drawbacks are discussed in the next section.

2.2 Advantages and Disadvantages of Ore Grade Estimation Models

Regression analysis is regarded as one of the most reliable approaches for estimating ore grade. It is a statistical approach for examining the relationships between variables (Mirjalili et al., 2020). Regression methods depend on the number of independent variables, the type of dependent variables, and the shape of the regression line (Panahi et al., 2021a). The assumption of linearity cannot provide an accurate model when there is a nonlinear relationship.

Similarly, a regression may over fit if it starts modeling the data's random error (noise) (Fung et al., 2020).

The structure of artificial neural networks is flexible, nonlinear, and similar to a nerve system (Kapageridis, 2005; Ehteram et al., 2022; Hanoon et al., 2021). An ANN has the main advantage: no procedure or process must be defined between inputs and outputs (Bazrafshan et al., 2022; Panahi et al., 2021b). If the data size is small, the model performance can be unsatisfactory, and if the data size is large, the model becomes complex (Ehteram et al., 2021).

Input spaces can be converted to output spaces using fuzzy logic. Fuzzy logic can model imprecise data and nonlinear functions (Fung et al., 2020). When the number of fuzzy rules increases, computational costs increases (Mirjalili et al., 2020). A support vector machine model (SVM) aims to reduce the generalization error, while an ANN attempts to minimize the training error. Researchers can develop a robust SVM model by tuning the parameters and using different kernels (Huang et al., 2022). The limitation of the choice of kernels is one of the disadvantages of SVM models (Fung et al., 2020). A hybrid model is an evolutionary machine learning model. Evolutionary machine learning models are created when an optimization algorithm is coupled with a machine learning model. Optimization algorithms enhance machine learning accuracy (Panahi et al., 2021b; Samanta et al., 2006). They can find the model parameters accurately.

2.3 Shortcomings of Previous Studies

While the previous studies developed a model for estimating ore grade, they had shortcomings:

1. The previous studies used individual machine learning models. An individual machine learning model includes one model, such as an ANN, SVM, or regression analysis model (Ahmadi et al., 2022; Emami et al., 2022). Ensemble models include multiple individual models. Using multiple models can improve the accuracy of outputs.
2. Previous studies did not consider the uncertainty of model parameters and inputs. Inputs and model parameters represent uncertainties. As a result, modelers must consider the effect of uncertainty resources on their outputs.
3. Previous studies have not reduced the computational time of machine learning models. Models are time-consuming, which is one of their disadvantages.
4. Previous studies did not introduce an effective method for choosing the best input scenario.

This book addresses these shortcomings by developing and introducing new models for estimating ore grades.

2.4 General Remarks

This chapter reviewed the application of machine learning algorithms for estimating ore grades. Previous papers demonstrated that machine learning models could accurately estimate ore grade. However, each model has its advantages and disadvantages. In addition, the previous papers had limitations. Thus, it is necessary to develop new models to address the shortcomings of previous papers. The ANN model was one of the most successful models for estimating ore grade. Model performance depends on the number of data points and the model parameters. Robust algorithms are required to determine ANN model parameters. The computational time of ANN models is another key factor. Modelers should reduce the computational time of ANN models. The other issue is to consider the uncertainty resources. The ANNs are also individual. Individual ANN models need to be improved by using an ensemble model. The next chapter tries to develop ANN models based on different strategies for these reasons.

References

Abbaszadeh, M., Ehteram, M., Ahmed, A. N., Singh, V. P., & Elshafie, A. (2021). The copper grade estimation of porphyry deposits using machine learning algorithms and henry gas solubility optimization. *Earth Science Informatics*. https://doi.org/10.1007/s12145-021-00667-6

Abuntori, C. A., Al-Hassan, S., & Mireku-Gyimah, D. (2021). Assessment of ore grade estimation methods for structurally controlled vein deposits - a review. *Ghana Mining Journal, 21*(1), 31–44. https://doi.org/10.4314/gm.v21i1.4

Ahmadi, A., Olyaei, M., Heydari, Z., Emami, M., Zeynolabedin, A., Ghomlaghi, A., & Sadegh, M. (2022). Groundwater level modeling with machine learning: A systematic review and meta-analysis. *Water, 14*(6), 949.

Bazdar, H., Fattahi, H., & Ghadimi, F. (2015). Hybrid ANN with invasive weed optimization algorithm, a new technique for prediction of gold and silver in zarshuran gold deposit, Iran. *Quarterly Journal of Tethys, 3*(3), 273–285.

Bazrafshan, O., Ehteram, M., Latif, S. D., Huang, Y. F., Teo, F. Y., Ahmed, A. N., & El-Shafie, A. (2022). Predicting crop yields using a new robust Bayesian averaging model based on multiple hybrid ANFIS and MLP models. *Ain Shams Engineering Journal, 13*(5), 101724.

Chatterjee, S., Bandopadhyay, S., Ganguli, R., Bhattacherjee, A., Samanta, B., & Pal, S. K. (2007). General regression neural network residual estimation for ore grade prediction of limestone deposit. *Transactions of the Institutions of Mining and Metallurgy, Section A: Mining Technology, 116*(3), 89–99. https://doi.org/10.1179/174328607X228875

Chatterjee, S., Bandopadhyay, S., & Rai, P. (2008). Genetic algorithm-based neural network learning parameter selection for ore grade evaluation of limestone deposit. *Transactions of the Institutions of Mining and Metallurgy, Section A: Mining Technology, 117*(4), 178–190. https://doi.org/10.1179/037178409X405732

Chatterjee, S., Bandopadhyay, S., & Machuca, D. (2010). Ore grade prediction using a genetic algorithm and clustering based ensemble neural network model. *Mathematical Geosciences, 42*(3), 309–326. https://doi.org/10.1007/s11004-010-9264-y

Dutta, S., Bandopadhyay, S., Ganguli, R., & Misra, D. (2010). Machine learning algorithms and their application to ore reserve estimation of sparse and imprecise data. *Journal of Intelligent Learning Systems and Applications, 02*(02), 86–96. https://doi.org/10.4236/jilsa.2010.22012

Ehteram, M., Ahmed, A. N., Kumar, P., Sherif, M., & El-Shafie, A. (2021). Predicting freshwater production and energy consumption in a seawater greenhouse based on ensemble frameworks using optimized multi-layer perceptron. *Energy Reports, 7*, 6308–6326.

Ehteram, M., Panahi, F., Ahmed, A. N., Huang, Y. F., Kumar, P., & Elshafie, A. (2022). Predicting evaporation with optimized artificial neural network using multi-objective salp swarm algorithm. *Environmental Science and Pollution Research*. https://doi.org/10.1007/s11356-021-16301-3.

Emami, M., Ahmadi, A., Daccache, A., Nazif, S., Mousavi, S. F., & Karami, H. (2022). County-level irrigation water demand estimation using machine learning: Case study of California. *Water, 14*(12), 1937.

Fung, K. F., Huang, Y. F., Koo, C. H., & Soh, Y. W. (2020). Drought forecasting: A review of modelling approaches 2007–2017. *Journal of Water and Climate Change, 11*(3), 771–799. https://doi.org/10.2166/wcc.2019.236

Hanoon, M. S., Ahmed, A. N., Zaini, N. A., Razzaq, A., Kumar, P., Sherif, M., & El-Shafie, A. (2021). Developing machine learning algorithms for meteorological temperature and humidity forecasting at Terengganu state in Malaysia. *Scientific Reports, 11*(1), 1–19.

Huang, H., Band, S. S., Karami, H., Ehteram, M., Chau, K. W., & Zhang, Q. (2022). Solar radiation prediction using improved soft computing models for semi-arid, slightly-arid and humid climates. *Alexandria Engineering Journal, 61*(12), 10631–10657.

Jafrasteh, B., & Fathianpour, N. (2017a). Optimal location of additional exploratory drillholes using afuzzy-artificial bee colony algorithm. *Arabian Journal of Geosciences, 10*(9), 1–16. https://doi.org/10.1007/s12517-017-2948-6

Jafrasteh, B., & Fathianpour, N. (2017b). A hybrid simultaneous perturbation artificial bee colony and back-propagation algorithm for training a local linear radial basis neural network on ore grade estimation. *Neurocomputing*. https://doi.org/10.1016/j.neucom.2017.01.016

Kapageridis, I. (2005). Input space configuration effects in neural network-based grade estimation. *Computers & Geosciences, 31*(6), 704–717. https://doi.org/10.1016/j.cageo.2005.01.001

Kaplan, U. E., & Topal, E. (2020). A new ore grade estimation using combine machine learning algorithms. *Minerals, 10*(10), 847.

Karami, R., & Afzal, P. (2015). Estimation of elemental distributions by combining artificial neural network and inverse distance weighted (IDW) based on lithogeochemical data in kahang porphry deposit, Central Iran. *Universal Journal of Geoscience, 3*(2), 59–65. https://doi.org/10.13189/ujg.2015.030203

Koike, K., & Matsuda, S. (2003). Characterizing content distributions of impurities in a limestone mine using a feedforward neural network. *Natural Resources Research, 12*(3), 209–222. https://doi.org/10.1023/A:1025180005454

Majdifar, S., & Kamali, G. (2014). Iron grade estimation using ANFIS algorithm at tappeghermez anomaly of sangan mine. *Journal of Analytical and Numerical Methods in Mining Engineering (ANM),* (5). [in Persian]

Mirjalili, S., Faris, H., & Aljarah, I. (2020). *Introduction to evolutionary machine learning techniques* (pp. 1–7). Springer, Singapore. https://doi.org/10.1007/978-981-32-9990-0_1

Panahi, F., Ahmed, A. N., Singh, V. P., Ehtearm, M., elshafie, A., & Haghighi, A. T. (2021a). Predicting freshwater production in seawater greenhouses using hybrid artificial neural network models. *Journal of Cleaner Production.* https://doi.org/10.1016/j.jclepro.2021.129721

Panahi, F., Ehteram, M., Ahmed, A. N., Huang, Y. F., Mosavi, A., & El-Shafie, A. (2021b). Streamflow prediction with large climate indices using several hybrid multilayer perceptrons and copula Bayesian model averaging. *Ecological Indicators.* https://doi.org/10.1016/j.ecolind.2021.108285

Patel, A. K., Chatterjee, S., & Gorai, A. K. (2017). Development of machine vision-based ore classification model using support vector machine (SVM) algorithm. *Arabian Journal of Geosciences, 10*(5), 1–16. https://doi.org/10.1007/s12517-017-2909-0

Samanta, B., Bandopadhyay, S., & Ganguli, R. (2002). Data segmentation and genetic algorithms for sparse data division in nome placer gold grade estimation using neural network and geostatistics. *Exploration and Mining Geology, 11*, 69–76. https://doi.org/10.2113/11.1-4.69

Samanta, B., Banopadhyay, S., Ganguli, R., & Dutta, S. (2005a). A comparative study of the performance of single neural network vs. Adaboost algorithm based combination of multiple neural networks for mineral resource estimation. *Journal of the Southern African Institute of Mining and Metallurgy, 105*(4), 237–246.

Samanta, B., Ganguli, R., & Bandopadhyay, S. (2005b). Comparing the predictive performance of neural networks with ordinary kriging in a bauxite deposit. *Mining Technology, 114*(3), 129–139. https://doi.org/10.1179/037178405X53980

Samanta, B., Bandopadhyay, S., & Ganguli, R. (2006). Comparative evaluation of neural network learning algorithms for ore grade estimation. *Mathematical Geosciences, 38*, 175–197. https://doi.org/10.1007/s11004-005-9010-z

Soltani-Mohammadi, S., Hoseinian, F. S., Abbaszadeh, M., & Khodadadzadeh, M. (2022). Grade estimation using a hybrid method of back-propagation artificial neural network and particle swarm optimization with integrated samples coordinate and local variability. *Computers and Geosciences, 159*, 104981. https://doi.org/10.1016/j.cageo.2021.104981

Tahmasebi, P., & Hezarkhani, A. (2012). A hybrid neural networks-fuzzy logic-genetic algorithm for grade estimation. *Computers & Geosciences, 42*, 18–27. https://doi.org/10.1016/j.cageo.2012.02.004.

Wu, X., & Zhou, Y. (1993). Reserve estimation using neural network techniques. *Computers & Geosciences, 19*(4), 567–575. https://doi.org/10.1016/0098-3004(93)90082-G.

Chapter 3
Structure of Different Kinds of ANN Models

Abstract This chapter discusses how artificial neural networks (ANNs) can be used to estimate ore grades. ANN models can successfully estimate ore grades using advanced operators. The advantages of each ANN model are discussed in this chapter. There are different layers in ANN models. Layers perform different tasks. The performance of ANN models depends on the parameters of ANNs. Different ANN models are compared for estimating ore grade in this chapter. A modeler can choose the best ANN model by understanding its different features.

Keywords ANNs · Ore grade · Predictive models · Radial basis function neural network models

3.1 Introduction

The ANN is one of the most important soft computing models (Seifi et al., 2022; Chong et al., 2023). The advantages of ANN models are their simplicity, low cost, and high efficiency (Seifi et al., 2021). Because of its simplicity and practicality, the Feedforward Neural Network (FFNN) is the most popular type of neural network (Banadkooki et al., 2020). FFNNs receive input and produce output by connecting neurons at different layers.

FFNNs can be classified into two types: Single-Layer Perceptron (SLP) and Multi-Layer Perceptron (MLP). SLPs can be used for linear problems because of their single perceptron. An MLP has multiple hidden neurons arranged in different layers. Consequently, it can solve nonlinear problems.

The radial basis function neural network (RBFNN) is one of the most important neural networks. An RBF network has a single hidden layer and Gaussian activation functions and is a multi-layer feedforward neural network (Yaseen et al., 2019). The advantages of RBF networks are fast learning speeds, easy parameter adjustment, and strong generalization abilities. The structure and the model parameters are two major components of RBF networks, like other neural networks (Ehteram et al., 2021a).

For single-hidden-layer feedforward neural networks (SFLNs), a new neural learning algorithm, called extreme learning machine (ELM), was developed to

M. Ehteram et al., *Estimating Ore Grade Using Evolutionary Machine Learning Models*, https://doi.org/10.1007/978-981-19-8106-7_3

improve their performance (Ding et al., 2014). ELMs consist of learning parameters, such as input weights and biases, which are randomly assigned, while the output weights are obtained by applying generalized inverse operation. In SLFNs, ELM provides faster learning than other traditional learning algorithms (Sattar et al., 2019).

ANN has a sub-model called group method of data handling (GMDH). GMDH employs a family of inductive algorithms to model multiparameter data sets using computers (Amiri & Soleimani, 2021; Shen et al., 2019).

MLP, RBFNN, GMDH, and ELM are the most famous ANN models. Thus, this book focuses on applying these models for estimating ore grades. For this reason, this book estimates ore grades using these models. Additionally, these models will be developed in the next chapters. The next section reviews the applications of these models for different fields.

3.1.1 A Review of Studies of Applying MLP, RBFNN, GMDH, and ELM for Estimating Different Variables in Geosciences and Mining Engineering, and Other Fields

- **Geological Mapping**
  Machin learning models (MLMs) can be used for supervised geological mapping using airborne geophysics and remotely sensed data (Leverington, 2010). The SVM (Cracknell & Reading, 2014; Oommen et al., 2008; Shebl & Csámer, 2021; Yu et al., 2012), ANN (Cracknell & Reading, 2014; Leverington, 2010; Shebl & Csámer, 2021), Maximum Likelihood Classifier (Shebl & Csámer, 2021), Naive Bayes, k-Nearest Neighbors (Cracknell & Reading, 2014), random forest (Albert & Ammar, 2021; Cracknell & Reading, 2014; Kuhn et al., 2018), and ensemble models (Waske & Braun, 2009) are widely used for Geological mapping.

- **Mineral Potential Mapping**
  Mineral potential mapping aims to identify areas with high mineral potential. Data-driven and knowledge-driven methods can be used for mineral potential mapping. The ANNs (Leite & de Souza Filho, 2009a; Saljoughi & Hezarkhani, 2018; Skabar, 2007) and support vector machines (Abedi et al., 2012; Rodriguez-Galiano et al., 2014, 2015; Saljoughi & Hezarkhani, 2018; Zuo & Carranza, 2011) are the most important data-driven methods.

- **Geochemical Data Processing**
  In geochemical data processing, MLMs are widely used to detect anomalies. Because of the Complexity of geological data, traditional and statistical methods cannot detect geochemical anomalies. Thus, some MLMs have been applied to identify geochemical anomalies (Gonbadi et al., 2015; Twarakavi et al., 2006).

We can use these techniques without making assumptions about data distribution (Zhang et al., 2022). Various ML methods have been used for anomaly detection, such as support vector machines (Gonbadi et al., 2015; Twarakavi et al., 2006), artificial neural networks (Beucher et al., 2013; Yu et al., 2019; Ziaii et al., 2012), random forest (Gonbadi et al., 2015; Wang et al., 2019a, 2019b), continuum-limited Boltzmann machine (Chen et al., 2014), metric learning (Wang et al., 2019a, 2019b), and autoencoder approaches (Chen et al., 2019a, 2019b; Luo et al., 2020; Xiong & Zuo, 2022).

- **Geomechanics and Geotechnics**
  Various MLMs are applied to estimate or classify the geomechanical and geotechnical parameters, such as estimation of modulus of elasticity (Acar & Kaya, 2020; Banan et al., 2020; Shahani et al., 2022), uniaxial compressive strength (Banan et al., 2020; Barzegar et al., 2020; Kim et al., 2020; Özdemir, 2022), unconfined compressive strength (Gowida et al., 2021; Jahed Armaghani et al., 2015), shear strength (Zhou et al., 2019), Young's modulus (Jahed Armaghani et al., 2015), porosity (Behnia et al., 2017; Gamal & Elkatatny, 2021; Gamal et al., 2021), the ground vibration caused by blasting operation (Bakhshandeh Amnieh et al., 2012; Ghasemi et al., 2016; Mohammadi et al., 2011; Soltani-Mohammadi et al., 2012), and bearing capacity of the cohesive soft soils (Bunawan et al., 2018).

- **Water Resource Management**
  Akiner (2021) estimated rainfall using ANN models. A two-layer feedforward ANN was trained using the Levenberg–Marquardt algorithm (LMA). The results showed that the ANN model had a high potential for predicting rainfall. Moosavi et al. (2021) combined GMDH with signal processing techniques to develop a powerful model for forecasting groundwater levels. GMDH was reported as a promising model to forecast groundwater levels. Alizamir et al. (2018) investigated the potential of the ELM to predict groundwater levels. For predicting groundwater levels, the ELM outperformed ANN models.

- **Energy**
  Khosravi et al. (2018) used GMDH to predict the time-series wind speed data. GMDH accurately predicted the wind speed data for all time intervals. Hai et al. (2020) used ELM to estimate global incident solar radiation (GISR). For predicting GISR, the ELM outperformed multiple linear regressions. Li et al. (2022) used radial basis neural networks (RBFNNs) and a metaheuristic algorithm to predict power. Based on the results, the optimized RBFNN was beneficial for practical applications.

- **Landslide and Flood**
  Wang et al. (2021) used MLP models for predicting floods. The MLP model was the most effective among the six methods in their study. Tiwari et al. (2021) used ANN, ELM, and M5 model trees to predict floods. ELM and M5 model trees performed better than ANN models. Elkurdy et al. (2021) used the GMDH model to predict floods. The results showed that the GMDH had a high potential for predicting floods in complex boundary condition. Huang et al. (2022) used RBFNN models to identify landslide susceptibility zones. Based on the results, the RBFNN model performed the best.

- **Ore Grade Estimation**
  Kaplan and Topal (2020) estimated ore grade using the RBFNN model. The models were created based on geological information (lithology and alteration) and drill hole data (easting, northing, and altitude). It was found that the RBFNN model had high accuracy for estimating ore grade. Fathi et al. (2021) used an ELM model to predict ore grade using borehole location, depth, and drill hole information. The ELM was highly capable of predicting ore grades. Hajkarimian et al. (2022) used optimized MLP models to estimate ore grade. They found that the optimization algorithms improved the performance of MLP models for estimating ore grade. MLP, ELM, RBFNN, and GMDH have been applied successfully for estimating ore grade and other target variables. The next sections will explain the structure of each model.

3.2 Structure of Multi-Layer Perceptron Models

Each layer of MLP plays a role in the modeling process. Input data enters the first layer (Liang et al., 2021; Panahi et al., 2021; Emami et al., 2022). Middle layers are called hidden layers. In the hidden layer, input data are processed (Ehteram et al., 2021a). The last layer gives the final output. Each neuron in MLPs produces output based on the summation and activation functions. First, using Eq. (3.1), we calculate the weighted sum (Ehteram et al., 2021b).

$$Z_j = \sum_{i=1}^{n} \omega_{ij} I_i + \beta_j \tag{3.1}$$

where Z_j: the weighted sum, ω_{ij}: weight connection input and hidden layer, β_j: bias, and I_i: input, and n: number of input variables. The second step is to apply the activation function using the output of neurons, as shown in Eq. (3.1). Several kinds of activation functions can be used for training MLPs. Previous studies proved the sigmoid function to be the optimal activation function for MLPs (Ehteram et al., 2021b).

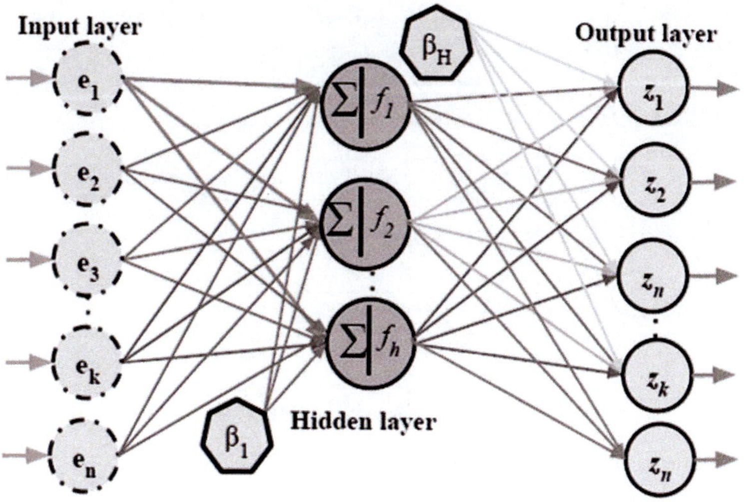

Fig. 3.1 Structure of the MLP model

$$f_j(x) = \frac{1}{1 + e^{-Z_j}} \tag{3.2}$$

As a result, the output of the MLP is given in Eq. (3.3):

$$out_i = f_j\left(\sum_{i=1}^{n} \omega_{ij} I_i + \beta_j\right) \tag{3.3}$$

where out_i: final output. Figure 3.1 shows the structure of the MLP model.

3.3 Structure of RBFNN Models

The RBF network can be described as a three-layer feedforward neural network (Hanoon et al., 2021). A Gaussian function is commonly used as an activation function in RBFNs. Receiving input data is the responsibility of the first layer (He et al., 2019). Activation functions are used to process data in the hidden layer. In RBF, there is a single hidden layer, while MLP may have several hidden layers. An RBFNN model has a nonlinear hidden layer and a linear output layer (Darabi et al., 2021). An RBFNN's activation function is based on the distance between the pattern and the prototype.

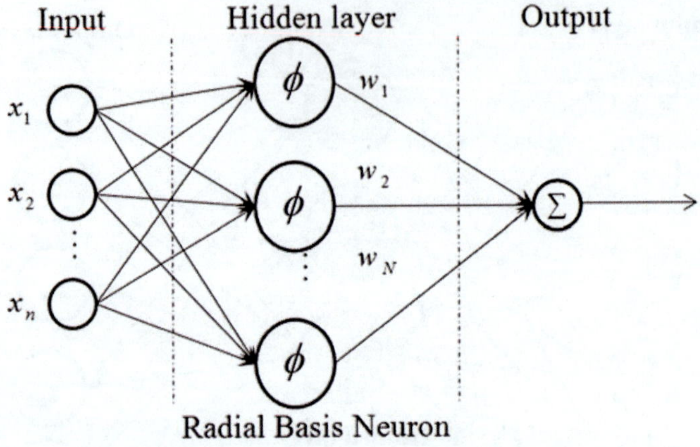

Fig. 3.2 Structure of the RBFNN model

$$\mu_i(\|x - c_i\|) = \exp\left(-\frac{\|x - c_i\|^2}{2\sigma^2}\right) \tag{3.4}$$

where x: input, c_i: the center of hidden neuron, σ the neuron's width, and output of the ith activation function.

$$\text{out}_k = \sum_{j=1}^{n} \omega_{jk}\mu_i(x) \tag{3.5}$$

where out_k: final output, ω_{jk}: weight connection. Figure 3.2 shows the structure of the RBFNN model (He et al., 2019).

3.4 Structure of Extreme Learning Machine (ELM) Models

HUANG et al. (2006) introduced the ELM model. ELM is faster than conventional algorithms such as ANN and SVM. The ELM is a learning algorithm for feedforward neural networks with one hidden layer (single-hidden-layer feedforward neural networks (SFLNs)).

$$\text{out}_j = \sum_{i=1}^{N} \varphi_i f_i(x_j) = \sum_{i=1}^{N} \varphi_i f(x_j; \alpha_i, v_i) \tag{3.6}$$

where out_j: the output vector, $\alpha_i = [\alpha_{i,1}, \alpha_{i,2}, .., \alpha_{i,n}]$, v_i: learning parameters $\varphi_i = [\varphi_{i,1}, \varphi_{i,2}, .., \phi_{i,m}]^T$: the weight connection linking the jth hidden node and the output nodes, and $f(x_j; \alpha_i, v_i)$: the activation function of ELM. Set $\alpha_i . xj$ be the inner product of α_i and xj. Equation (3.6) can be rewritten as (Ding et al., 2014):

$$H\varphi = Out \tag{3.7}$$

where $H = \begin{bmatrix} f(\alpha_1 x_1 + v_1) & & f(\alpha_N x + v_N) \\ . & . & . \\ f(\alpha_1 x_N + v_1) & & f(\alpha_N x_N + v_N) \end{bmatrix}, \varphi = \begin{bmatrix} \varphi_1^T \\ . \\ . \\ \varphi_N^T \end{bmatrix}$ and $O = \begin{bmatrix} out_1^T \\ \\ out_n^T \end{bmatrix}$.

In ELM theory, the hidden nodes' learning parameters α_i and v_i are randomly selected without considering the input data to minimize the network cost function. Therefore, Eq. (3.7) becomes a linear system, and a least square solution can be found for determining the output weights.

$$\hat{\varphi} = H^{\dagger} T \tag{3.8}$$

where $\hat{\varphi}$ the optimal solution $H^{\dagger}$: Moore–Penrose generalized inverse of H, H: The output matrix of the hidden layer. First, a training data set, an activation function, and the hidden node number were determined (Ding et al., 2014). Then, the parameters of hidden nodes were randomly selected. Next, we compute the output matrix of the hidden layer (Ding et al., 2014). Figure 3.3 shows the structure of the ELM model (Zhang et al., 2017).

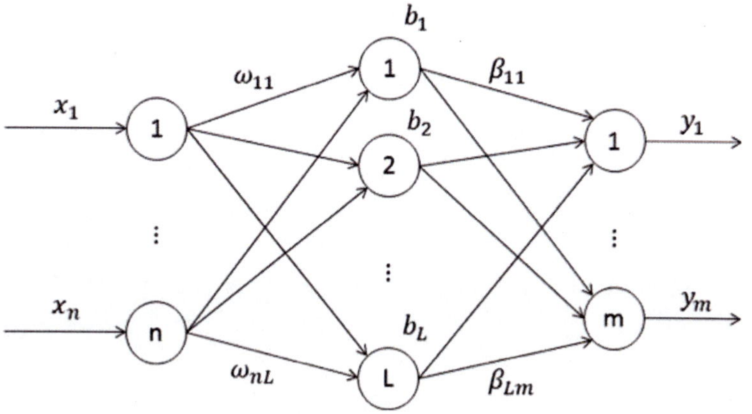

Fig. 3.3 Structure of the ELM model

3.5 Structure of Group Method of Data Handling Neural Networks

Based on the principles of self-organized learning, GMDH is used to model nonlinear systems. The GMDH approximates the relationship between inputs and outputs using nonlinear mapping. In GMDHs, the interconnections between neurons in successive layers are not necessarily fully interconnected (Amiri & Soleimani, 2021). While GMDH utilizes a polynomial transfer function, MLPNN may use different types of activation functions. There is no need to manually determine the model parameters. (Amiri & Soleimani, 2021). The mathematical model of GMDH is as follows:

$$o = b_0 + \sum_{i=1}^{m} b_i x_i + \sum_{i=1}^{m}\sum_{j=1}^{m} b_{ij} x_{ij} + \sum_{i=1}^{m}\sum_{j=1}^{m}\sum_{k=1}^{m} b_{ij} x_i x_j x_k \tag{3.9}$$

where o: output, x_i, x_j, and x_k: input, m: number of input variables. Equation (3.9) is converted to the following equation:

$$o = b_0 + b_1 x_1 + b_2 x_2 + b_3 x_1^2 + b_4 x_2^2 + b_5 x_1 x_2 \tag{3.10}$$

Based on the following regression system, GMDH calculates the weight coefficients vector (b):

$$b = \left(A^T A\right)^{-1} A^T O \tag{3.11}$$

$$A = \begin{bmatrix} 1\ x_1^1\ x_2^1\ x_1^1 x_2^1\ \left(x_1^1\right)^2\ \left(x_2^1\right)^2 \\ 1\ x_2^1\ x_2^2\ x_1^2 x_2^2\ \left(x_2^1\right)^2\ \left(x_2^2\right)^2 \\ \\ \\ 1\ x_m^1 \qquad\qquad\qquad \left(x_m^2\right)^2 \end{bmatrix} \tag{3.12}$$

where O: output matrix. The number of the next layer neurons of GMDH can be computed as

$$N_n = \binom{N_{np}}{2} \tag{3.13}$$

where N_{np}: the number of neurons of the current layer. The number of neurons in each layer is limited using the selection pressure criterion to prevent excessive complexity.

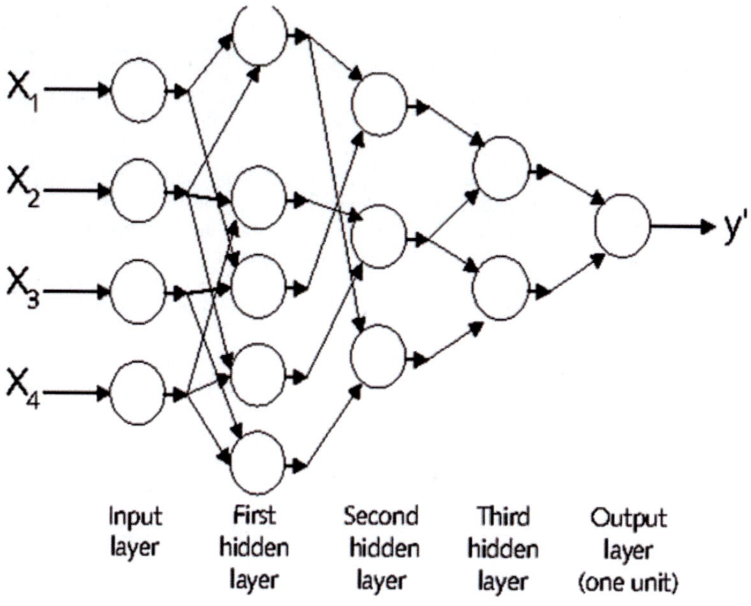

Fig. 3.4 Structure of the GMDH model

$$e_C = W \times \mathrm{RMSE}_{\mathrm{best}} + (1 - W) \times \mathrm{RMSE}_{\mathrm{worst}} \tag{3.14}$$

where $\mathrm{RMSE}_{\mathrm{best}}$: RMSE for the best neuron, and $\mathrm{RMSE}_{\mathrm{worst}}$: RMSE for the worst neuron, W: election pressure parameter. Figure 3.4 shows the structure of the GMDH model (Schrijver et al., 2002).

3.6 General Remarks

This chapter discussed famous ANN models such as MLP, RBFNN, ELM, and GMDH. The structures of the models were different. These models have unknown parameters. As a result, it is necessary to determine the values of these parameters for modeling. Advanced optimization algorithms and training algorithms can be used to develop these models. The best results can be obtained by adjusting the model parameters. Each of the models has advantages. The RBFNN models can have faster convergence than the MLP model.

ELM requires more hidden neurons. The main advantages of GMDH are its accuracy and flexibility. On the other hand, the modelers select a suitable model based on their target.

References

Abedi, M., Norouzi, G. H., & Bahroudi, A. (2012). Support vector machine for multi-classification of mineral prospectivity areas. *Computers & Geosciences, 46*, 272–283. https://doi.org/10.1016/J.CAGEO.2011.12.014

Acar, M. C., & Kaya, B. (2020). Models to estimate the elastic modulus of weak rocks based on least square support vector machine. *Arabian Journal of Geosciences, 13*(14), 1–12.

Akiner, M. E. (2021). Long-term rainfall information forecast by utilizing constrained amount of observation through artificial neural network approach. In *Advances in Meteorology*.

Albert, G., & Ammar, S. (2021). Application of random forest classification and remotely sensed data in geological mapping on the Jebel Meloussi area (Tunisia). *Arabian Journal of Geosciences, 14*(21), 2240. https://doi.org/10.1007/s12517-021-08509-x

Alizamir, M., Kisi, O., & Zounemat-Kermani, M. (2018). Modelling long-term groundwater fluctuations by extreme learning machine using hydro-climatic data. *Hydrological Sciences Journal, 63*(1), 63–73. https://doi.org/10.1080/02626667.2017.1410891

Amiri, M., & Soleimani, S. (2021). ML-based group method of data handling: An improvement on the conventional GMDH. *Complex & Intelligent Systems, 7*(6), 2949–2960. https://doi.org/10.1007/s40747-021-00480-0

Amnieh, H. B., Siamaki, A., & Soltani, S. (2012). Design of blasting pattern in proportion to the peak particle velocity (PPV): Artificial neural networks approach. *Safety Science, 50*(9). https://doi.org/10.1016/j.ssci.2012.05.008

Banadkooki, F. B., Ehteram, M., Ahmed, A. N., Teo, F. Y., Ebrahimi, M., Fai, C. M., Huang, Y. F., & El-Shafie, A. (2020). Suspended sediment load prediction using artificial neural network and ant lion optimization algorithm. *Environmental Science and Pollution Research, 27*(30), 38094–38116. https://doi.org/10.1007/s11356-020-09876-w

Banan, A., Nasiri, A., & Taheri-Garavand, A. (2020). Deep learning-based appearance features extraction for automated carp species identification. *Aquacultural Engineering, 89*, 102053. https://doi.org/10.1016/j.aquaeng.2020.102053

Barzegar, R., Sattarpour, M., Deo, R., Fijani, E., & Adamowski, J. (2020). An ensemble tree-based machine learning model for predicting the uniaxial compressive strength of travertine rocks. *Neural Computing and Applications, 32*(13), 9065–9080.

Behnia, D., Ahangari, K., & Moeinossadat, S. R. (2017). Modeling of shear wave velocity in limestone by soft computing methods. *International Journal of Mining Science and Technology, 27*(3), 423–430. https://doi.org/10.1016/j.ijmst.2017.03.006

Beucher, A., Österholm, P., Martinkauppi, A., Edén, P., & Fröjdö, S. (2013). Artificial neural network for acid sulfate soil mapping: Application to the sirppujoki river catchment area, south-western Finland. *Journal of Geochemical Exploration, 125*, 46–55. https://doi.org/10.1016/J.GEXPLO.2012.11.002

Bunawan, A. R., Momeni, E., Armaghani, D. J., & Rashid, A. S. A. (2018). Experimental and intelligent techniques to estimate bearing capacity of cohesive soft soils reinforced with soil-cement columns. *Measurement, 124*, 529–538.

Chen, Y., Lu, L., & Li, X. (2014). Application of continuous restricted Boltzmann machine to identify multivariate geochemical anomaly. *Journal of Geochemical Exploration, 140*, 56–63. https://doi.org/10.1016/J.GEXPLO.2014.02.013

Chen, L., Guan, Q., Feng, B., Yue, H., Wang, J., & Zhang, F. (2019a). A multi-convolutional autoencoder approach to multivariate geochemical anomaly recognition. In *Minerals* (Vol. 9, Issue 5). https://doi.org/10.3390/min9050270

Chen, L., Guan, Q., Xiong, Y., Liang, J., Wang, Y., & Xu, Y. (2019b). A spatially constrained multi-autoencoder approach for multivariate geochemical anomaly recognition. *Computers & Geosciences, 125*, 43–54. https://doi.org/10.1016/J.CAGEO.2019.01.016

Chong, K. L., Huang, Y. F., Koo, C. H., Sherif, M., Ahmed, A. N., & El-Shafie, A. (2023). Investigation of cross-entropy-based streamflow forecasting through an efficient interpretable automated search process. *Applied Water Science, 13*(1), 1–32.

Cracknell, M. J., & Reading, A. M. (2014). Geological mapping using remote sensing data: A comparison of five machine learning algorithms, their response to variations in the spatial distribution of training data and the use of explicit spatial information. *Computers & Geosciences, 63*, 22–33. https://doi.org/10.1016/J.CAGEO.2013.10.008

Darabi, H., Mohamadi, S., Karimidastenaei, Z., Kisi, O., Ehteram, M., ELShafie, A., & Torabi Haghighi, A. (2021). Prediction of daily suspended sediment load (SSL) using new optimization algorithms and soft computing models. *Soft Computing, 25*(11), 7609–7626.

Ding, S., Xu, X., & Nie, R. (2014). Extreme learning machine and its applications. In *Neural Computing and Applications* (Vol. 25, Issues 3–4, pp. 549–556). Springer. https://doi.org/10.1007/s00521-013-1522-8

Ehteram, M., Ahmed, A. N., Latif, S. D., Huang, Y. F., Alizamir, M., Kisi, O., Mert, C., & El-Shafie, A. (2021a). Design of a hybrid ANN multi-objective whale algorithm for suspended sediment load prediction. *Environmental Science and Pollution Research, 28*(2), 1596–1611. https://doi.org/10.1007/s11356-020-10421-y

Ehteram, M., Sammen, S. S., Panahi, F., & Sidek, L. M. (2021b). A hybrid novel SVM model for predicting CO2 emissions using Multiobjective Seagull Optimization. *Environmental Science and Pollution Research, 28*(46), 66171–66192.

Elkurdy, M., Binns, A. D., Bonakdari, H., Gharabaghi, B., & McBean, E. (2021). Early detection of riverine flooding events using the group method of data handling for the Bow River, Alberta, Canada. *International Journal of River Basin Management*. https://doi.org/10.1080/15715124.2021.1906261

Emami, M., Ahmadi, A., Daccache, A., Nazif, S., Mousavi, S. F., & Karami, H. (2022). County-level irrigation water demand estimation using machine learning: Case study of California. *Water, 14*(12), 1937.

Fathi, M., Alimoradi, A., & Ahooi, H. H. (2021). Optimizing extreme learning machine algorithm using particle swarm optimization to estimate iron ore grade. *Journal of Mining and Environment, 12*(2), 397–411. https://doi.org/10.22044/jme.2021.10368.1984

Gamal, H., & Elkatatny, S. (2021). Prediction model based on an artificial neural network for rock porosity. *Arabian Journal for Science and Engineering*. https://doi.org/10.1007/s13369-021-05912-0

Gamal, H., Elkatatny, S., Alsaihati, A., & Abdulraheem, A. (2021). Intelligent prediction for rock porosity while drilling complex lithology in real time. *Computational Intelligence and Neuroscience, 2021*, 9960478. https://doi.org/10.1155/2021/9960478

Ghasemi, E., Kalhori, H., & Bagherpour, R. (2016). A new hybrid ANFIS–PSO model for prediction of peak particle velocity due to bench blasting. *Engineering with Computers, 32*(4), 607–614.

Gonbadi, A. M., Tabatabaei, S. H., & Carranza, E. J. M. (2015). Supervised geochemical anomaly detection by pattern recognition. *Journal of Geochemical Exploration, 157*, 81–91. https://doi.org/10.1016/J.GEXPLO.2015.06.001

Gowida, A., Elkatatny, S., & Gamal, H. (2021). Unconfined compressive strength (UCS) prediction in real-time while drilling using artificial intelligence tools. *Neural Computing and Applications, 33*(13), 8043–8054.

Hai, T., Sharafati, A., Mohammed, A., Salih, S. Q., Deo, R. C., Al-Ansari, N., & Yaseen, Z. M. (2020). Global solar radiation estimation and climatic variability analysis using extreme learning machine based predictive model. *IEEE Access, 8*, 12026–12042. https://doi.org/10.1109/ACCESS.2020.2965303

Hajkarimian, H., Alimoradi, A., Hemati Ahooi, H., & Salsabili, M. (2022). Comparison between the performance of four metaheuristic algorithms in training a multilayer perceptron machine for gold grade estimation. *International Journal of Mining and Geo-Engineering, 0*. https://doi.org/10.22059/IJMGE.2021.314154.594880

Hanoon, M. S., Ahmed, A. N., Zaini, N. A., Razzaq, A., Kumar, P., Sherif, M., & El-Shafie, A. (2021). Developing machine learning algorithms for meteorological temperature and humidity forecasting at Terengganu state in Malaysia. *Scientific Reports, 11*(1), 1–19.

He, H., Yan, Y., Chen, T., & Cheng, P. (2019). Tree height estimation of forest plantation in mountainous terrain from bare-earth points using a DoG-coupled radial basis function neural network. *Remote Sensing, 11*(11), 1271.

Huang, G. B., Chen, L., & Siew, C. K. (2006). Universal approximation using incremental constructive feedforward networks with random hidden nodes. *IEEE Transactions on Neural Networks, 17*(4), 879–892. https://doi.org/10.1109/TNN.2006.875977

Huang, J., Ling, S., Wu, X., & Deng, R. (2022). GIS-based comparative study of the Bayesian network, decision table, radial basis function network and stochastic gradient descent for the spatial prediction of landslide susceptibility. *Land, 11*(3), 436. https://doi.org/10.3390/land11030436

Jahed Armaghani, D., Tonnizam Mohamad, E., Momeni, E., & Narayanasamy, M. S. (2015). An adaptive neuro-fuzzy inference system for predicting unconfined compressive strength and Young's modulus: a study on Main Range granite. *Bulletin of Engineering Geology and the Environment, 74*(4), 1301–1319.

Kaplan, U. E., & Topal, E. (2020). A new ore grade estimation using combine machine learning algorithms. *Minerals, 10*(10), 847.

Khosravi, A., Machado, L., & Nunes, R. O. (2018). Time-series prediction of wind speed using machine learning algorithms: A case study Osorio wind farm, brazil. *Applied Energy, 224*, 550–566. https://doi.org/10.1016/j.apenergy.2018.05.043

Kim, T.-H., Ko, T. Y., Park, Y. S., Kim, T. K., & Lee, D. H. (2020). Prediction of uniaxial compressive strength of rock using shield TBM machine data and machine learning technique. *Tunnel and Underground Space, 30*(3), 214–225.

Kuhn, S., Cracknell, M. J., & Reading, A. M. (2018). Lithologic mapping using random forests applied to geophysical and remote-sensing data: A demonstration study from the eastern goldfields of Australia. *Geophysics, 83*(4), B183–B193. https://doi.org/10.1190/geo2017-0590.1

Leite, E. P., & de Souza Filho, C. R. (2009a). Probabilistic neural networks applied to mineral potential mapping for platinum group elements in the Serra Leste region, Carajás Mineral Province, Brazil. *Computers & Geosciences, 35*(3), 675–687. https://doi.org/10.1016/j.cageo.2008.05.003

Leverington, D. W. (2010). Discrimination of sedimentary lithologies using Hyperion and Landsat thematic mapper data: A case study at Melville Island, Canadian High Arctic. *International Journal of Remote Sensing, 31*(1), 233–260.

Li, T., Liu, X., Lin, Z., & Morrison, R. (2022). Ensemble offshore wind turbine power curve modelling – an integration of isolation forest, fast radial basis function neural network, and metaheuristic algorithm. *Energy, 239*, 122340. https://doi.org/10.1016/j.energy.2021.122340

Liang, G., Panahi, F., Ahmed, A. N., Ehteram, M., Band, S. S., & Elshafie, A. (2021). Predicting municipal solid waste using a coupled artificial neural network with archimedes optimisation algorithm and socioeconomic components. *Journal of Cleaner Production, 315*, 128039.

Luo, Z., Xiong, Y., & Zuo, R. (2020). Recognition of geochemical anomalies using a deep variational autoencoder network. *Applied Geochemistry, 122*, 104710. https://doi.org/10.1016/J.APGEOCHEM.2020.104710

Mohammadi, S. S., Amnieh, H. B., & Bahadori, M. (2011). Predicting ground vibration caused by blasting operations in sarcheshmeh copper mine considering the charge type by adaptive neuro-fuzzy inference system (ANFIS). *Archives of Mining Sciences, 56*(4).

Moosavi, V., Mahjoobi, J., & Hayatzadeh, M. (2021). Combining group method of data handling with signal processing approaches to improve accuracy of groundwater level modeling. *Natural Resources Research, 30*(2), 1735–1754. https://doi.org/10.1007/s11053-020-09799-w

Oommen, T., Misra, D., Twarakavi, N. K. C., Prakash, A., Sahoo, B., & Bandopadhyay, S. (2008). An objective analysis of support vector machine based classification for remote sensing. *Mathematical Geosciences, 40*(4), 409–424.

Özdemir, E. (2022). A new predictive model for uniaxial compressive strength of rock using machine learning method: Artificial intelligence-based age-layered population structure genetic programming (ALPS-GP). *Arabian Journal for Science and Engineering, 47*(1), 629–639.

Panahi, F., Ehteram, M., & Emami, M. (2021). Suspended sediment load prediction based on soft computing models and Black Widow Optimization Algorithm using an enhanced gamma test. *Environmental Science and Pollution Research, 28*(35), 48253–48273.

Rodriguez-Galiano, V. F., Chica-Olmo, M., & Chica-Rivas, M. (2014). Predictive modelling of gold potential with the integration of multisource information based on random forest: a case study on the Rodalquilar area, Southern Spain. *International Journal of Geographical Information Science, 28*(7), 1336–1354. https://doi.org/10.1080/13658816.2014.885527

Rodriguez-Galiano, V., Sanchez-Castillo, M., Chica-Olmo, M., & Chica-Rivas, M. (2015). Machine learning predictive models for mineral prospectivity: An evaluation of neural networks, random forest, regression trees and support vector machines. *Ore Geology Reviews, 71*, 804–818. https://doi.org/10.1016/j.oregeorev.2015.01.001

Saljoughi, B. S., & Hezarkhani, A. (2018). A comparative analysis of artificial neural network (ANN), wavelet neural network (WNN), and support vector machine (SVM) data-driven models to mineral potential mapping for copper mineralizations in the Shahr-e-Babak region, Kerman, Iran. *Applied Geomatics, 10*(3), 229–256. https://doi.org/10.1007/s12518-018-0229-z

Sattar, A. M. A., Ertuğrul, Ö. F., Gharabaghi, B., McBean, E. A., & Cao, J. (2019). Extreme learning machine model for water network management. *Neural Computing and Applications, 31*(1), 157–169. https://doi.org/10.1007/s00521-017-2987-7

Schrijver, M., Kerckhoffs, E., Water, P., & Saman, K. (2002). Model tuning with the use of heuristic-free GMDH (group method of data handling) networks. In *14th European Simulation Symposium*.

Seifi, A., Ehteram, M., Nayebloei, F., Soroush, F., Gharabaghi, B., & Torabi Haghighi, A. (2021). GLUE uncertainty analysis of hybrid models for predicting hourly soil temperature and application wavelet coherence analysis for correlation with meteorological variables. *Soft Computing, 25*(16), 10723–10748.

Seifi, A., Ehteram, M., Soroush, F., & Haghighi, A. T. (2022). Multi-model ensemble prediction of pan evaporation based on the Copula Bayesian Model Averaging approach. *Engineering Applications of Artificial Intelligence, 114*, 105124.

Shahani, N. M., Zheng, X., Guo, X., & Wei, X. (2022). Machine learning-based intelligent prediction of elastic modulus of rocks at thar coalfield. *Sustainability, 14*(6), 3689.

Shebl, A., & Csámer, Á. (2021). Stacked vector multi-source lithologic classification utilizing machine learning algorithms: Data potentiality and dimensionality monitoring. *Remote Sensing Applications: Society and Environment, 24*, 100643. https://doi.org/10.1016/J.RSASE.2021.100643

Shen, C., Asante-Okyere, S., Ziggah, Y. Y., Wang, L., & Zhu, X. (2019). Group method of data handling (GMDH) lithology identification based on wavelet analysis and dimensionality reduction as well log data pre-processing techniques. *Energies, 12*(8), 1509. https://doi.org/10.3390/en12081509

Skabar, A. (2007). Mineral potential mapping using Bayesian learning for multilayer perceptrons. *Mathematical Geology, 39*(5), 439–451. https://doi.org/10.1007/s11004-007-9106-8

Soltani-Mohammadi, S., Amnieh, H. B., & Bahadori, M. (2012). Investigating ground vibration to calculate the permissible charge weight for blasting operations of gotvand-olya dam underground structures. *Archives of Mining Sciences, 57*(3). https://doi.org/10.2478/v10267-012-0044-z

Tiwari, M. K., Deo, R. C., & Adamowski, J. F. (2021). Short-term flood forecasting using artificial neural networks, extreme learning machines, and M5 model tree. In *Advances in Streamflow Forecasting* (pp. 263–279). Elsevier. https://doi.org/10.1016/b978-0-12-820673-7.00012-3

Twarakavi, N. K. C., Misra, D., & Bandopadhyay, S. (2006). Prediction of arsenic in bedrock derived stream sediments at a gold mine site under conditions of sparse data. *Natural Resources Research, 15*(1), 15–26. https://doi.org/10.1007/s11053-006-9013-6

Wang, Z., Dong, Y., & Zuo, R. (2019a). Mapping geochemical anomalies related to Fe–polymetallic mineralization using the maximum margin metric learning method. *Ore Geology Reviews, 107*, 258–265. https://doi.org/10.1016/J.OREGEOREV.2019.02.027

Wang, Z., Zuo, R., & Dong, Y. (2019b). Mapping geochemical anomalies through integrating random forest and metric learning methods. *Natural Resources Research, 28*(4), 1285–1298. https://doi.org/10.1007/s11053-019-09471-y

Wang, Y., Fang, Z., Hong, H., Costache, R., & Tang, X. (2021). Flood susceptibility mapping by integrating frequency ratio and index of entropy with multilayer perceptron and classification and regression tree. *Journal of Environmental Management, 289*, 112449. https://doi.org/10.1016/j.jenvman.2021.112449

Waske, B., & Braun, M. (2009). Classifier ensembles for land cover mapping using multitemporal SAR imagery. *ISPRS Journal of Photogrammetry and Remote Sensing, 64*(5), 450–457.

Xiong, Y., & Zuo, R. (2022). Robust feature extraction for geochemical anomaly recognition using a stacked convolutional denoising autoencoder. *Mathematical Geosciences, 54*(3), 623–644. https://doi.org/10.1007/s11004-021-09935-z

Yaseen, Z. M., Sulaiman, S. O., Deo, R. C., & Chau, K. W. (2019). An enhanced extreme learning machine model for river flow forecasting: State-of-the-art, practical applications in water resource engineering area and future research direction. *Journal of Hydrology, 569*, 387–408. https://doi.org/10.1016/j.jhydrol.2018.11.069

Yu, L., Porwal, A., Holden, E.-J., & Dentith, M. C. (2012). Towards automatic lithological classification from remote sensing data using support vector machines. *Computers & Geosciences, 45*, 229–239.

Yu, X., Xiao, F., Zhou, Y., Wang, Y., & Wang, K. (2019). Application of hierarchical clustering, singularity mapping, and kohonen neural network to identify Ag-Au-Pb-Zn polymetallic mineralization associated geochemical anomaly in pangxidong district. *Journal of Geochemical Exploration, 203*, 87–95. https://doi.org/10.1016/J.GEXPLO.2019.04.007

Zhang, S., Zhang, T., Yin, Y., & Xiao, W. (2017). Alumina concentration detection based on the kernel extreme learning machine. *Sensors (switzerland), 17*(9), 2002. https://doi.org/10.3390/s17092002

Zhang, H., Wang, J., Wang, M., & Chen, X. (2022). Integration of cuckoo search and fuzzy support vector machine for intelligent diagnosis of production process quality. *Journal of Industrial & Management Optimization, 18*(1), 195–217. https://doi.org/10.3934/jimo.2020150

Zhou, J., Li, E., Wei, H., Li, C., Qiao, Q., & Armaghani, D. J. (2019). Random forests and cubist algorithms for predicting shear strengths of rockfill materials. *Applied Sciences, 9*(8), 1621.

Ziaii, M., Doulati Ardejani, F., Ziaei, M., & Soleymani, A. A. (2012). Neuro-fuzzy modeling based genetic algorithms for identification of geochemical anomalies in mining geochemistry. *Applied Geochemistry, 27*(3), 663–676. https://doi.org/10.1016/J.APGEOCHEM.2011.12.020

Zuo, R., & Carranza, E. J. M. (2011). Support vector machine: A tool for mapping mineral prospectivity. *Computers & Geosciences, 37*(12), 1967–1975.

Chapter 4
Optimization Algorithms and Classical Training Algorithms

Abstract This chapter explains the application of different optimization algorithms and training algorithms for training artificial neural network models (ANN). The different optimization algorithms are reliable tools for training ANN models. These algorithms use advanced operators to train the ANN models. The ANN parameters, such as bias and weight, are unknown. Thus, robust optimization algorithms can be used for adjusting ANN parameters. This chapter explains the structure of different optimization algorithms for solving optimization problems. The advantages and disadvantages of different optimization algorithms are explained in this chapter. The optimization algorithms use advanced operators to adjust the ANN parameters.

Keywords Optimization algorithms · Training algorithms · ANN models · ANN parameters

4.1 Introduction

Several artificial neural network models (ANNs) were discussed in the previous section. Each of the models has unknown parameters. Identifying the unknown parameters of models is essential before starting the modeling process. Modeling accuracy is affected by the values of ANN parameters, which must be optimized. Using classical and optimization algorithms, model parameters can be calculated. This chapter will explain several classical training algorithms, including backpropagation, scaled conjugate gradient, and Levenberg–Marquardt algorithms. These algorithms allow us to find unknown model parameters, but they may be trapped in the local optimum. First, we will discuss classical training algorithms. Next, the structure of optimization algorithms is described. Lastly, we will discuss the various structures of evolutionary artificial neural networks.

M. Ehteram et al., *Estimating Ore Grade Using Evolutionary Machine Learning Models*, https://doi.org/10.1007/978-981-19-8106-7_4

4.1.1 Backpropagation Algorithm

Training algorithms influence the accuracy of ANN models. Training algorithms are applied to find the optimal weights and biases (Yaghi et al., 2020; Pham et al., 2020). A backpropagation algorithm is a gradient-based approach. Backpropagation algorithms have two phases: the forward phase and the backward phase. The forward phase involves finding the network parameters and propagating the input data (Ahmadian, 2016). The output error is calculated in the forward phase. The estimated error is propagated through the network in the backward step (Ahmadian, 2016). A network parameter is adjusted to minimize error. Backpropagation algorithms suffer from slow convergence.

4.2 Levenberg–Marquardt Algorithm (LM)

The nonlinear least squares problem is solved using the LM. Using the LM method, the Hessian matrix (*H*) is not necessary to be calculated (Ali et al., 2022). In such cases, the Jacobian matrix (*J*) can be used to approximate the Hessian matrix (Ali et al., 2022).

$$H = J^T J \tag{4.1}$$

The iteration process based on the following equation is used to update the ANN parameters:

$$w_{\mathrm{k}+1} = w_k - \left[J^T J + \mu I\right]^{-1} J^T e \tag{4.2}$$

where $w_{\mathrm{k}+1}$: : the weight connection at iteration k + 1, w_k: the weight connection at iteration *k*, *e*: error value, *I*: identity matrix, and μ: scalar.

4.3 Scaled Conjugate Gradient Algorithm

The gradient conjugate algorithm (CGA) uses conjugate directions but does not perform a line search at each iteration (Rezaeian-Zadeh et al., 2013). As a first step, conjugate gradient algorithms look for the steepest descent direction:

$$P_0 = -g_0 \tag{4.3}$$

where P_0: initial weight vectors and g_0: the initial gradient. An optimal distance is then determined based on a line search along the current direction of the search:

$$X_{K+1} = X_K + \alpha_K . P_K \tag{4.4}$$

where X_{K+1}: the new weight vector value, X_K the vector of current weights and biases, α_K and the learning rate. Combining the new steepest descent direction with the previous search direction generally determines the new search direction

$$P_k = -g_k + \beta_k . P_{k-1} \tag{4.5}$$

where g_k: the current gradient and β_k: constant coefficient.

4.4 Variable Learning Rate Algorithm

The learning rate remains constant when the steepest descent is used for training. The gradient descent is as follows (Rezaeian-Zadeh et al., 2013):

$$x_{k+1} = x_k - \alpha_k g_k \tag{4.6}$$

where x_{k+1}: a vector of new weights and biases, g_k: current gradient, and α_k: learning rate. The algorithm's accuracy depends on how the learning rate is set. The algorithm can oscillate and become unstable when the learning rate is increased. Convergence of the algorithm will be slow if the learning rate is too low.

4.5 Optimization Algorithm

For solving real-world problems, search methods were the first artificial intelligence technique. Since the 1940s, many algorithms have been developed to solve optimization problems (Emami et al., 2021). There are two categories of search methods: uninformed and informed methods (Ehteram et al., 2017; Panahi et al., 2021b; Banadkooki et al., 2020). An uninformed search algorithm has no information about the state or the search space. However, these methods are not practical for computationally expensive problems, as they are time-consuming (Tikhamarine et al., 2020). The second group uses optimization algorithms for solving problems.

By using such heuristics, we can achieve a higher speed (Valikhan-Anaraki et al., 2019). The structure of famous optimization algorithms is explained in this chapter. The next chapter will explain how different artificial neural networks can be trained using optimization algorithms. Fast convergence, high efficiency, flexibility, and accuracy are the advantages of these algorithms.

4.5.1 Salp Swarm Algorithm (SSA)

The SSA was introduced by Mirjalili et al. (2017). It mimics the behavior of Salps in the ocean. The SSA has been used in different fields. Faris et al. (2018) tested the newly proposed SSA. The proposed method outperformed other algorithms. A chaos SSA was proposed by Sayed et al. (2018). Based on the results, chaotic logistic maps could significantly improve the original SSA's performance. The chaotic version of SSA was proposed by Ahmed et al. (2018). The proposed approach addresses feature selection problems. By using chaotic maps, the SSA algorithm was significantly improved.

Abusnaina et al. (2018) suggested SSA for optimizing neural network parameters. This method outperformed other algorithms. Ekinci and Hekimoğlu (2018) used SSA in multimachine power systems to tune the power system stabilizer (PSS). SSA performed better than other intelligent approaches. A binary version of the SSA (BSSA) was proposed by Rizk-Allah et al. (2019). The modified BSSA performed better than other variants. SSA was modified and applied by Qais et al. (2019). Results showed that the enhanced SSA improved convergence. Tubishat et al. (2021) introduced an improved version of SSA called the Dynamic Salp Swarm Algorithm (DSSA). The DSSA algorithm outperformed other algorithms. Singh et al. (2022) combined SSA with PSO to remove unprofitable characteristics. The proposed hybrid method often provided superior results than other existing algorithms. Salgotra et al. (2021) introduced new versions of SSA. Results demonstrated that the proposed SSA (MSSA) was the best algorithm among other algorithms.

A salp belongs to the Salpidae family and has a transparent body. Their tissues are similar to jellyfish. Like jellyfish, salps also move forward by pumping water through their bodies. Salp chains divide into two groups: leaders and followers. Leaders are at the front of the chain, and followers follow them. Followers follow the leader, and the leader guides the swarm. In the search space, there is also a food source. The following equation is proposed to update the position of the leader:

$$\mathrm{le}_j^l = \begin{bmatrix} F_j + \chi_1\left((\mathrm{up}_i - \mathrm{lo}_i)\chi_2 + \mathrm{lo}_i\right) \leftarrow \chi_3 \geq 0 \\ F_j - \chi_1\left((\mathrm{up}_i - \mathrm{lo}_i)\chi_2 + \mathrm{lo}_i\right) \leftarrow \chi_3 < 0 \end{bmatrix} \tag{4.7a}$$

where le_j^l: the location of the leader, up_i: upper bound of variable, lo_i: lower bound of variable, χ_1, χ_3, and χ_2: random values, and F_j: location of food. The coefficient χ_1 balances exploration and exploitation in SSA, so it is the most important parameter.

$$\chi_1 = 2e^{-\left(\frac{4l}{L}\right)} \tag{4.7b}$$

where l: number of the current iteration and L: maximum number of iterations. The followers update their location based on the following equation:

$$\mathrm{salp}_j^i = \frac{1}{2}\left(\mathrm{salp}_j^i + \mathrm{salp}_j^{i-1}\right) \tag{4.8}$$

where $salp'^{i}_{j}$: the location of ith salp in jth dimension and $salp'^{i-1}_{j}$: the location of ith salp in jth dimension. Figure 4.1 shows the structure of SSA. The advantages of algorithms are fast convergence and ease of implementation.

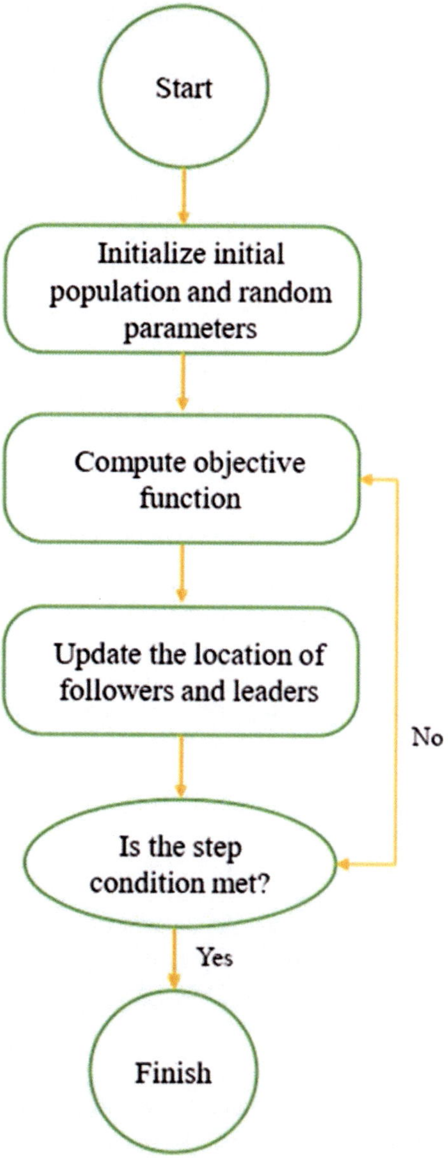

Fig. 4.1 Structure of salp swarm algorithm

4.5.2 Sine Cosine Algorithm (SCA)

Mirjalili (2016) suggested a new population-based optimization algorithm called SCA to solve optimization problems. In the modeling process, sine and cosine functions are used (Khozani et al., 2022; Seifi et al., 2021). Attia et al. (2018) solved the optimal power flow (OPF) problem based on modified SCA. For the OPF problem under different objective functions, MSCA provides a simple and robust solution. Abd Elaziz et al. (2017) developed a version of SCA using opposition-based learning (OBL). The proposed method found optimal solutions in complex search spaces efficiently. Li et al. (2018) found SVM parameters using SCA. The proposed model could optimally determine SVR parameters. Using a novel enhanced SCA, Gupta and Deep (2019) improved the exploitation abilities of SCA. The improved SCA could be used to solve optimization problems efficiently. Using the elite candidate solution, Gupta et al. (2020) modified the classical search equation of the SCA. Results showed that the modified SCA was more efficient than other algorithms. Ullah et al. (2020) solved the energy optimization problem. The SCA efficiently minimized energy consumption. Tubishat et al. (2022) used a chaotic map and the Simulated Annealing (SA) algorithm to improve the SCA. Results confirmed that hybrid SCA outperformed other optimization algorithms (Mirjalili, 2016). The SCA avoids local optima by creating random solutions and improving them. The SCA algorithm begins the optimization process based on random solutions. Once all solutions have been determined, the algorithm assigns the best solution to the destination point and updates all the other solutions (Mirjalili, 2016). When the number of iterations exceeds the maximum value, the SCA algorithm terminates the optimization process (Achite et al., 2022).

The SCA algorithm smoothly switches between exploration and exploitation using sine and cosine functions (Mirjalili, 2016). The SCA updates its solution based on the following equation:

$$S_i^{t+1} = \begin{bmatrix} S_i^t + r_1 \times \sin(r_2) \times \left(r_3 P_i^t - S_i^t\right) \leftarrow r_4 < 0.50 \\ S_i^t + r_1 \times \cos(r_2) \times \left(r_3 P_i^t - S_i^t\right) \leftarrow r_4 \geq 0.50 \end{bmatrix} \tag{4.9}$$

where S_i^{t+1}: : new solution at iteration t + 1, P_i^t: the best solution, r_1, r_2, r_3, and r_4: random parameters. The r_1 is updated as follows:

$$r_1 = a - \frac{i}{i_{\max}} a \tag{4.10}$$

where a: constant-coefficient, i: current iteration, and $i_{\max}$: maximum iteration numbers. Figure 4.2 shows the structure of SCA.

Fig. 4.2 Structure of SCA

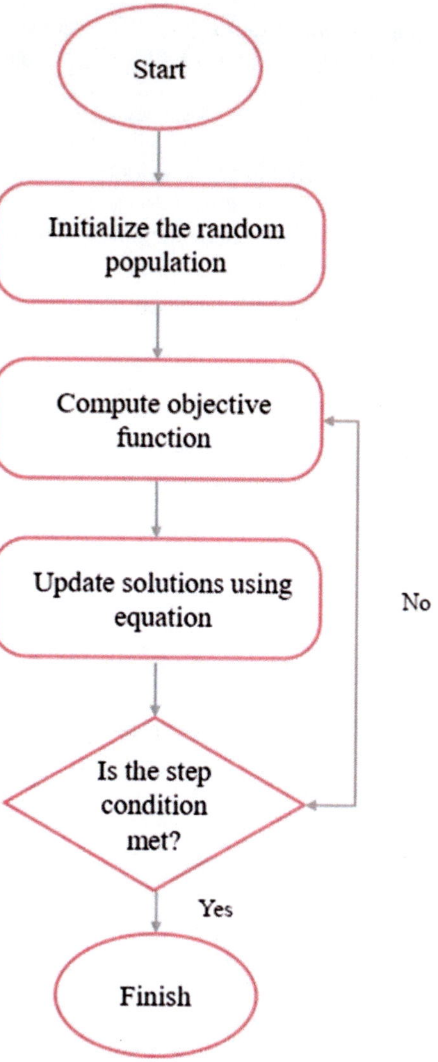

4.5.3 *Structure of Shark Swarm Optimization*

The Shark smell optimization (SSO) algorithm was introduced by Abedinia et al. (2016). Using neural networks and chaotic SSO, Abedinia and Amjady (2015) developed a hybrid wind power forecasting model. The Chaos SSO model was used to optimize the ANN parameters. The results showed that the SSO could find model parameters accurately. Seifi et al. (2020) used multiobjective SSO to train artificial neural networks. Results showed that multi-layer perceptron-SSO had the best ability among other models. Animals have an olfactory system that receives chemical signals from distant sources. Sharks use their smell sense to find the source of the odor. Sharks use their smell sensor to locate prey.

Mathematical formulations are constructed using some assumptions.

(1) The injured fish injects blood into the sea (the search area). Hence, it is considered prey for sharks.
(2) Blood is injected into seawater regularly. As concentration increases and becomes stronger, the shark increases its velocity. Sharks find their prey by smelling blood.

The initial locations of sharks are determined as follows:

$$\mathrm{sr}_i^J = \left[\mathrm{sr}_{i,1}^J, \mathrm{sr}_{i,2}^J, \ldots, \mathrm{sr}_{i,\mathrm{ND}}^J\right] \tag{4.11}$$

where sr_i^J: jth dimension of the shark's ith position, and ND: number of decision variables. Depending on the distance from the prey, the intensity of the odor changes at each position. Next, the shark's velocity is determined:

$$v_{i,j}^k = \varsigma_k.R_1.\frac{\partial(\mathrm{OF})}{\partial x_j}|_{x_{i,j}^k} + \alpha_k.R_2.v_{i,j}^{\mathrm{k}-1} \tag{4.12}$$

where $v_{i,j}^k$: the velocity of ith shark in tjh dimension, R_1: random number, ς_k: random number, α_k: rate of momentum or inertia coefficient, and OF: objective function. Sharks can increase their velocity up to a specific limit. For each stage of the SSO algorithm, a velocity limiter is used.

$$\left|v_{i,j}^k\right| = \min\left[\left|\varsigma_k.R_1.\frac{\partial(\mathrm{OF})}{\partial x_j}\right| + \alpha_K.R_2.v_{i,j}^{\mathrm{k}-1}|, \left|\beta_k, v_{i,j}^{\mathrm{k}-1}\right|\right] \tag{4.13}$$

where β_k: velocity limiter and R_2: random values. Since the shark is moving forward, its new position is as follows:

$$y_i^{\mathrm{k}+1} = x_i^k + v_i^k.\Delta t_k \tag{4.14}$$

where $y_i^{\mathrm{k}+1}$: the new location of the shark and Δt_k: interval time. Rotational movement is onc of the most important characteristics of sharks. The rotational movement of sharks is used to perform a local search. Sharks update their position by rotating:

$$z_i^{\mathrm{k}+1,m} = y_i^{\mathrm{k}+1} + R_3.y_i^{\mathrm{k}+1} \tag{4.15}$$

where $z_i^{\mathrm{k}+1,m}$: the location of the point m in the local search and M: number of points in the local search. Figure 4.3 shows the structure of SSO.

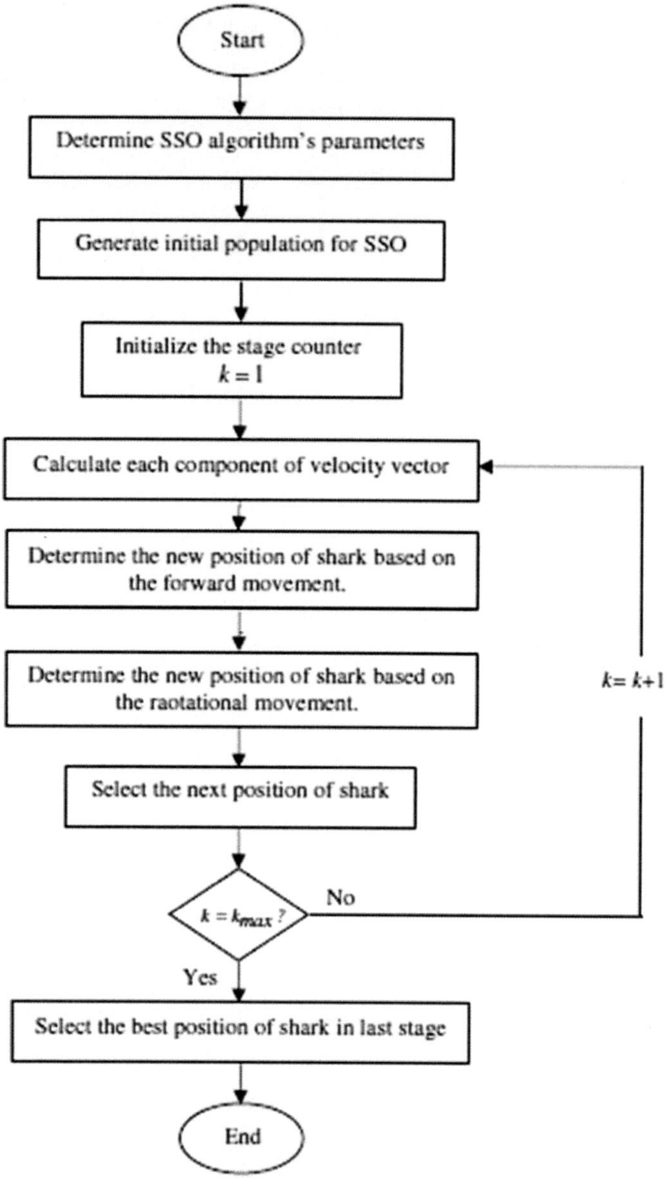

Fig. 4.3 Structure of SSO

4.5.4 Structure of Naked Mole-Rat (NMR) Algorithm

The NMR algorithm simulates NMR's mating patterns, and it has the following key features (Salgotra & Singh, 2019):

(1) NMRs live in social groups of 295 members, with an average of 70–80 members per group.
(2) Each group has a queen who divides the population into breeders and workers.
(3) A breeder is the best NMR. A queen mates with a breeder.
(4) The other workers perform other duties.
(5) The best workers perform all the necessary tasks and are replaced by the breeders.
(6) In a breeder pool, the queen mates with the best breeder.
(7) The first step is to initialize the population of the NMR.

$$\text{NM} = \text{NM}_{\min,j} + U(0,1) \times \left(\text{NM}_{\max,j} - \text{NM}_{\min,j}\right) \tag{4.16}$$

where $\text{NM}_{i,j}$: the ith NMR in jth dimension, $U(0,1)$: uniformly distributed number, $\text{NM}_{\max,j}$: the upper bound of variables, and $\text{NM}_{\min,j}$: the lower bound of variables. The fitness function values are computed to identify the workers and breeders (Salgotra & Singh, 2019).

The worker phase is the next level. A worker tries to become a breeder and mate with the queen. Using local information and experience, workers can create a new NMR solution. If the new NMR's fitness is better, the old solution is discarded, and the new one is memorized. The fitness function values of all the workers are remembered once they complete the search process.

A new solution can be produced from an old one using the following equation:

$$\text{Wo}_i^{t+1} = \text{Wo}_i^t + \lambda\left(\text{Wo}_j^t - W_k^t\right) \tag{4.17}$$

where Wo_i^t: the ith worker in the t iteration, Wo_i^{t+1}: the new solution, and λ: mating factor. The breeder phase is the next level. Breeders update themselves to remain a breeder and be selected for mating. The breeders who cannot update their fitness may be pushed into the workers' category.

$$\text{br}_i^{t+1} = (1-\lambda)\text{br}_i^t + \lambda\left(d - \text{br}_i^t\right) \tag{4.18}$$

where br_i^{t+1}: the ith bredder in t iteration, λ: mating frequency and d: the best solution. The algorithm identifies breeders and workers. An initial evaluation is conducted to determine the best breeder and worker. As the workers improve their fitness, they may be promoted to breeders.

A breeder must update its fitness to remain a breeder. A sterile breeder is pushed into the workers' category (Salgotra & Singh, 2019). The structure of NMR is shown in Fig. 4.4. Workers constantly improve themselves and explore new areas of the search space to become a member of a breeder group (Salgotra & Singh, 2019). The breeders are fitter than workers and carry out more intensive exploitations because they strive to enhance their fitness to mate with the queen.

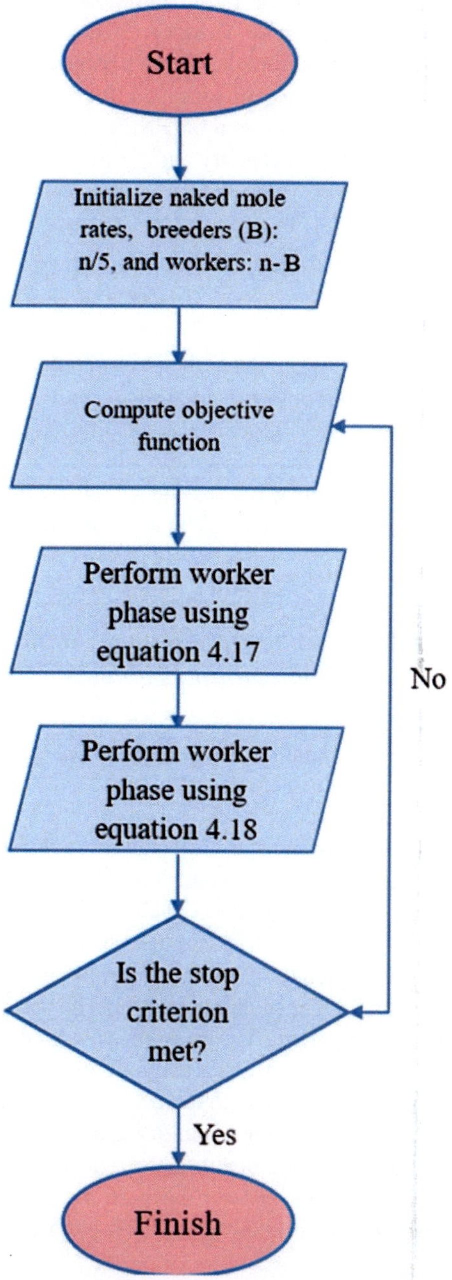

Fig. 4.4 Structure of proposed NMR for solving optimization problems

4.5.5 Structure of Particle Swarm Optimization

The particle swarm optimization (PSO) algorithm optimizes nonlinear continuous functions (Garro & Vázquez, 2015). It was inspired by bird flocks searching for food, which exhibited collective behavior. Faster convergence, easy implementation, and computation efficiency are the advantages of PSO (Garro & Vázquez, 2015). A particle's location in the PSO represents a decision variable's value. Particles are defined by their velocity and position:

$$\mathrm{ve}_i(\mathrm{t}+1) = \omega v_i(t) + \delta_1 r_1 (p_i(t) - x_i(t)) + \delta_2 r_1 (g_i(t) - x_i(t)) \tag{4.19}$$

$$x_i(\mathrm{t}+1) = x_i(t) + v_i(\mathrm{t}+1) \tag{4.20}$$

where $\mathrm{ve}_i(\mathrm{t}+1)$: the velocity of the ith particle at $\mathrm{t}+1$ iteration, $v_i(t)$: the velocity of the ith particle at t iteration, δ_1 and δ_2: acceleration coefficient, $p_i(t)$: local best solution, $g_i(t)$: global best location, $x_i(t)$: the location of ith solution at the tth iteration, r_1 and r_2: random numbers. During the system's iteration, particles in the swarm move toward the global optimum. A particle's velocity is affected by acceleration constants. The structure of PSO is shown in Fig. 4.5.

4.5.6 Structure of Genetic Algorithm for Solving Optimization Problems

GA is a population-based method. GA consists of chromosomes, and each chromosome contains genes.

Based on an objective function, the GA evaluates each member of the population. The best solutions are selected through random selection (e.g., the roulette wheel) to improve the poor solutions. The probability of choosing an optimal solution is proportional to the fitness value. It is possible to pull out good solutions trapped in a local optimum by using other solutions (Ehteram et al., 2021). This algorithm can predict a global optimum by maintaining and improving the best solutions in each generation. The mutation is another advantage of this algorithm (Panahi et al., 2021a). Changing genes in chromosomes randomly enhances the exploratory behavior of GA by maintaining diversity and individual differences. The GA algorithm begins with a randomly generated population. GA assigns probabilities to individuals and chooses them based on their suitability to become the next generation. A new chromosome is created by combining the genes of a male and a female. The crossover creates two new solutions. In the last step, we perform a mutation. After creating new solutions, a gene or multiple genes are altered. Through mutation, population diversity is preserved.

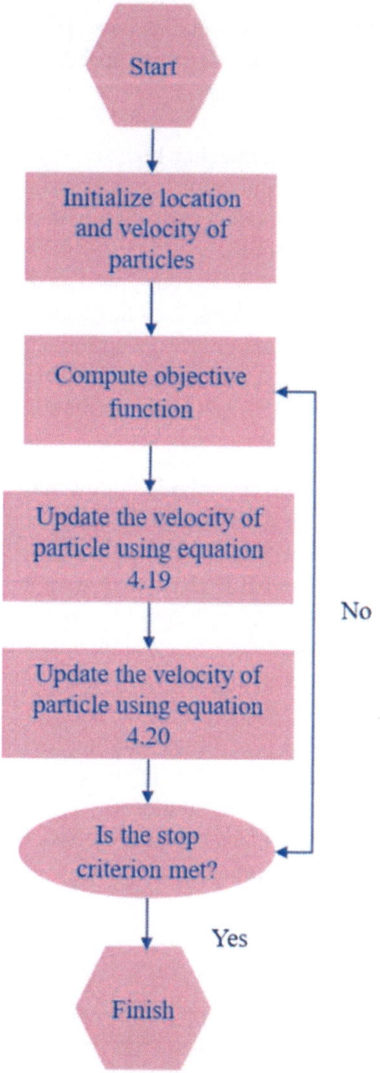

Fig. 4.5 Structure of proposed PSO for solving optimization problems

4.6 Evolutionary Multi-Layer Perceptron (MLP) and Radial Basis Function Neural Network Models (RBFNN)

MLP models contain unknown parameters such as weight connections and biases. These parameters can be considered decision variables. An optimization process should be defined to determine the optimal values of these parameters. An evolutionary MLP model combines an MLP and an optimization algorithm. An optimization algorithm is used as a training algorithm to determine the parameters of an

MLP. Previous studies have used evolutionary MLP models for different purposes. For non-stationary wind speed prediction, Liu et al. (2013) used GA-MLP and PSO-MLP. The GA and PSO significantly improved the accuracy of MLP. He et al. (2014) used MLP-PSO to predict particulate matter at the street level. For fine particles, the proposed model had high accuracy. Samadianfard et al. (2018) coupled MLP with the whale optimization algorithm to predict rainfall. MLP-WOA increased the accuracy of MLP models and was recommended for predicting annual rainfall. Pham et al. (2020) coupled intelligent water drop (IWD) with MLP to predict river streamflow. The MLP-IWD model produced better results than the classical MLP model and significantly improved the accuracy of river flow forecasting. Cai et al. (2021) used RBFNN and gray wolf optimization to predict soil liquefaction. They reported that the GWO could improve the efficiency of the RBFNN model.

The MLP and RBFNN models are combined with an optimization algorithm as follows:

(1) Data were divided into training and testing data.
(2) The MLP (RBFNN) model runs at the training level.
(3) Check the stop condition.
(4) The MLP (RBFNN) goes to the testing level if the stop condition is met; otherwise, it goes to step 5.
(5) MLP (RBFNN) parameters are defined as the initial population. An initial population of an MLP consists of the number of hidden neurons, hidden layers, biases, and weights.
(6) The algorithms' operators were used to update the positions of their agents.
(7) The objective function (root mean square error) is calculated to evaluate the quality of solutions.
(8) When the stop convergence is met, the process goes to step 3; otherwise, it goes to step 6.

4.7 Evolutionary Extreme Learning Machine (ELM)

One of the disadvantages of ELM is that its input weights are randomly selected. The random selection of input weights may result in inaccurate results. Optimization algorithms can be used to determine optimal input weights. Bui et al. (2019) performed a spatial prediction of flash floods using a PSO-ELM combination. The proposed model had a high ability to predict flash floods for high-frequency tropical typhoon areas. Cocco Mariani et al. (2019) coupled biography-based optimization to predict the conditions of a combustion parameter. The optimized ELM showed high potential for pressure prediction. The optimized ELM was generated based on the following levels:

(1) The data sets were divided into training and testing data.
(2) The ELM model runs at the training level.

(3) The stop condition is checked.
(4) If the stop condition is met, the ELM goes to the testing level; otherwise, the ELM goes to step 5.
(5) The input weights are defined as the initial population of algorithms.
(6) The RMSE is computed as an objective function.
(7) The input weights are the decision variables. Thus, the operators of algorithms are used to update the values of input weights.
(8) When the stop convergence is met, the process goes to step 4; otherwise, it goes to step 6.

4.8 Evolutionary Group Method of Data Handling Neural Networks (GMDH)

The performance of GMDH is dependent on the values of weight coefficients. Thus, it is necessary to determine the optimal values of weight coefficients. Mahdavi-Meymand and Zounemat-Kermani (2020) used an integrated GMDH-firefly algorithm (FFA) model to estimate air demand for spillway aerators. The performance of the standard GMDH was improved by using FFA.

Banik et al. (2022) combined GMDH with PSO to predict the flux of rectangular sheets. GMDH-PSO improved GMDH's accuracy. The initial values of weight coefficients are used to generate optimized GMDH models. Next, the GMDH models are run using training data. The RMSE is calculated as an objective function. Optimization algorithms are used to adjust the weights. A GMDH model moves to the testing level when the stop condition is met; otherwise, it returns to the starting point.

4.9 General Remarks

This section discusses how algorithms can improve the accuracy of artificial neural networks (ANNs). The classical training algorithms may converge to a local optimum. Each optimization algorithm has its advantages and disadvantages. Using advanced algorithms allows optimization algorithms to find the best solutions. The optimized ANN models will have a faster convergence rate than standalone models. Additionally, they will improve ANN models' accuracy. We will examine the performance of different ANNs in the next chapter.

References

Abd Elaziz, M., Oliva, D., & Xiong, S. (2017). An improved Opposition-Based Sine Cosine Algorithm for global optimization. *Expert Systems with Applications, 90*, 484–500. https://doi.org/10.1016/j.eswa.2017.07.043

Abedinia, O., & Amjady, N. (2015). Short-term wind power prediction based on hybrid neural network and chaotic shark smell optimization. *International Journal of Precision Engineering and Manufacturing - Green Technology, 2*(3), 245–254. https://doi.org/10.1007/s40684-015-0029-4

Abedinia, O., Amjady, N., & Ghasemi, A. (2016). A new metaheuristic algorithm based on shark smell optimization. *Complexity, 21*(5), 97–116. https://doi.org/10.1002/cplx.21634

Abusnaina, A. A., Ahmad, S., Jarrar, R., & Mafarja, M. (2018). Training neural networks using salp swarm algorithm for pattern classification. *ACM International Conference Proceeding Series, 10*(1145/3231053), 3231070.

Achite, M., Banadkooki, F. B., Ehteram, M., Bouharira, A., Ahmed, A. N., & Elshafie, A. (2022). Exploring Bayesian model averaging with multiple ANNs for meteorological drought forecasts. *Stochastic Environmental Research and Risk Assessment*, 1–26.

Ahmadian, A. S. (2016). *Numerical models for submerged breakwaters : Coastal hydrodynamics and morphodynamics.*

Ahmed, S., Mafarja, M., Faris, H., & Aljarah, I. (2018). Feature selection using salp swarm algorithm with chaos. In *ACM International Conference Proceeding Series*, 65–69. https://doi.org/10.1145/3206185.3206198

Ali, M. S., Ayaz, M., & Mansoor, T. (2022). Prediction of discharge through a sharp-crested triangular weir using ANN model trained with Levenberg–Marquardt algorithm. *Modeling Earth Systems and Environment, 8*(1), 1405–1417. https://doi.org/10.1007/s40808-021-01167-8

Attia, A. F., El Sehiemy, R. A., & Hasanien, H. M. (2018). Optimal power flow solution in power systems using a novel sine-cosine algorithm. *International Journal of Electrical Power and Energy Systems, 99*, 331–343. https://doi.org/10.1016/j.ijepes.2018.01.024

Banadkooki, F. B., Ehteram, M., Ahmed, A. N., Teo, F. Y., Ebrahimi, M., Fai, C. M., & El-Shafie, A. (2020). Suspended sediment load prediction using artificial neural network and ant lion optimization algorithm. *Environmental Science and Pollution Research, 27*(30), 38094–38116.

Banik, A., Majumder, M., Biswal, S. K., & Bandyopadhyay, T. K. (2022). Polynomial neural network-based group method of data handling algorithm coupled with modified particle swarm optimization to predict permeate flux (%) of rectangular sheet-shaped membrane. *Chemical Papers, 76*(1), 79–97. https://doi.org/10.1007/s11696-021-01838-x

Bui, D. T., Ngo, P. T. T., Pham, T. D., Jaafari, A., Minh, N. Q., Hoa, P. V., & Samui, P. (2019). A novel hybrid approach based on a swarm intelligence optimized extreme learning machine for flash flood susceptibility mapping. *CATENA, 179*, 184–196. https://doi.org/10.1016/j.catena.2019.04.009

Cai, M., Hocine, O., Mohammed, A. S., Chen, X., Amar, M. N., & Hasanipanah, M. (2021). Integrating the LSSVM and RBFNN models with three optimization algorithms to predict the soil liquefaction potential. *Engineering with Computers*, 1–13. https://doi.org/10.1007/s00366-021-01392-w

Cocco Mariani, V., Hennings Och, S., dos Santos Coelho, L., & Domingues, E. (2019). Pressure prediction of a spark ignition single cylinder engine using optimized extreme learning machine models. *Applied Energy, 249*, 204–221. https://doi.org/10.1016/j.apenergy.2019.04.126

di He, H., Lu, W. Z., & Xue, Y. (2014). Prediction of particulate matter at street level using artificial neural networks coupling with chaotic particle swarm optimization algorithm. *Building and Environment, 78*, 111–117. https://doi.org/10.1016/j.buildenv.2014.04.011

Ehteram, M., Ahmed, A. N., Kumar, P., Sherif, M., & El-Shafie, A. (2021). Predicting freshwater production and energy consumption in a seawater greenhouse based on ensemble frameworks using optimized multi-layer perceptron. *Energy Reports, 7*, 6308–6326. https://doi.org/10.1016/j.egyr.2021.09.079

Ehteram, M., Allawi, M. F., Karami, H., Mousavi, S. F., Emami, M., El-Shafie, A., & Farzin, S. (2017). Optimization of chain-reservoirs' operation with a new approach in artificial intelligence. *Water Resources Management, 31*(7), 2085–2104.

Ekinci, S., & Hekimoğlu, B. (2018). Parameter optimization of power system stabilizer via Salp Swarm algorithm. In *2018 5th International Conference on Electrical and Electronics Engineering, ICEEE 2018*, 143–147. https://doi.org/10.1109/ICEEE2.2018.8391318

Emami, M., Nazif, S., Mousavi, S. F., Karami, H., & Daccache, A. (2021). A hybrid constrained coral reefs optimization algorithm with machine learning for optimizing multi-reservoir systems operation. *Journal of Environmental Management, 286*, 112250.

Faris, H., Mafarja, M. M., Heidari, A. A., Aljarah, I., Al-Zoubi, A. M., Mirjalili, S., & Fujita, H. (2018). An efficient binary salp swarm algorithm with crossover scheme for feature selection problems. *Knowledge-Based Systems.* https://doi.org/10.1016/j.knosys.2018.05.009

Garro, B. A., & Vázquez, R. A. (2015). Designing artificial neural networks using particle swarm optimization algorithms. *Computational Intelligence and Neuroscience, 2015.* https://doi.org/10.1155/2015/369298

Gupta, S., & Deep, K. (2019). Improved sine cosine algorithm with crossover scheme for global optimization. *Knowledge-Based Systems, 165*, 374–406. https://doi.org/10.1016/j.knosys.2018.12.008

Gupta, S., Deep, K., Mirjalili, S., & Kim, J. H. (2020). A modified sine cosine algorithm with novel transition parameter and mutation operator for global optimization. *Expert Systems with Applications, 154*, 113395. https://doi.org/10.1016/j.eswa.2020.113395

Khozani, Z. S., Banadkooki, F. B., Ehteram, M., Ahmed, A. N., & El-Shafie, A. (2022). Combining autoregressive integrated moving average with long short-term memory neural network and optimisation algorithms for predicting ground water level. *Journal of Cleaner Production, 348*, 131224.

Li, S., Fang, H., & Liu, X. (2018). Parameter optimization of support vector regression based on sine cosine algorithm. *Expert Systems with Applications.* https://doi.org/10.1016/j.eswa.2017.08.038

Liu, H., Tian, H. Q., Chen, C., & Li, Y. F. (2013). An experimental investigation of two Wavelet-MLP hybrid frameworks for wind speed prediction using GA and PSO optimization. *International Journal of Electrical Power and Energy Systems, 52*(1), 161–173. https://doi.org/10.1016/j.ijepes.2013.03.034

Mahdavi-Meymand, A., & Zounemat-Kermani, M. (2020). A new integrated model of the group method of data handling and the firefly algorithm (GMDH-FA): Application to aeration modelling on spillways. *Artificial Intelligence Review, 53*(4), 2549–2569. https://doi.org/10.1007/s10462-019-09741-4

Mirjalili, S. (2016). SCA: A sine cosine algorithm for solving optimization problems. *Knowledge-Based Systems, 96*, 120–133. https://doi.org/10.1016/j.knosys.2015.12.022

Mirjalili, S., Gandomi, A. H., Mirjalili, S. Z., Saremi, S., Faris, H., & Mirjalili, S. M. (2017). Salp swarm algorithm: A bio-inspired optimizer for engineering design problems. *Advances in Engineering Software, 114*, 163–191. https://doi.org/10.1016/j.advengsoft.2017.07.002

Panahi, F., Ehteram, M., Ahmed, A. N., Huang, Y. F., Mosavi, A., & El-Shafie, A. (2021a). Streamflow prediction with large climate indices using several hybrid multilayer perceptrons and copula Bayesian model averaging. *Ecological Indicators, 133*, 108285. https://doi.org/10.1016/j.ecolind.2021.108285

Panahi, F., Ehteram, M., & Emami, M. (2021b). Suspended sediment load prediction based on soft computing models and Black Widow Optimization Algorithm using an enhanced gamma test. *Environmental Science and Pollution Research, 28*(35), 48253–48273.

Pham, Q. B., Afan, H. A., Mohammadi, B., Ahmed, A. N., Linh, N. T. T., Vo, N. D., & El-Shafie, A. (2020). Hybrid model to improve the river streamflow forecasting utilizing multi-layer perceptron-based intelligent water drop optimization algorithm. *Soft Computing, 24*(23), 18039–18056.

Qais, M. H., Hasanien, H. M., & Alghuwainem, S. (2019). Enhanced salp swarm algorithm: Application to variable speed wind generators. *Engineering Applications of Artificial Intelligence.* https://doi.org/10.1016/j.engappai.2019.01.011

Rezaeian-Zadeh, M., Tabari, H., & Abghari, H. (2013). Prediction of monthly discharge volume by different artificial neural network algorithms in semi-arid regions. *Arabian Journal of Geosciences, 6*(7), 2529–2537. https://doi.org/10.1007/s12517-011-0517-y

Rizk-Allah, R. M., Hassanien, A. E., Elhoseny, M., & Gunasekaran, M. (2019). A new binary salp swarm algorithm: development and application for optimization tasks. *Neural Computing and Applications, 31*(5), 1641–1663. https://doi.org/10.1007/s00521-018-3613-z

Salgotra, R., & Singh, U. (2019). The naked mole-rat algorithm. *Neural Computing and Applications, 31*(12), 8837–8857. https://doi.org/10.1007/s00521-019-04464-7

Salgotra, R., Singh, U., Singh, G., Singh, S., & Gandomi, A. H. (2021). Application of mutation operators to salp swarm algorithm. *Expert Systems with Applications.* https://doi.org/10.1016/j.eswa.2020.114368

Samadianfard, S., Ghorbani, M. A., & Mohammadi, B. (2018). Forecasting soil temperature at multiple-depth with a hybrid artificial neural network model coupled-hybrid firefly optimizer algorithm. *Information Processing in Agriculture, 5*(4), 465–476. https://doi.org/10.1016/j.inpa.2018.06.005

Sayed, G. I., Khoriba, G., & Haggag, M. H. (2018). A novel chaotic salp swarm algorithm for global optimization and feature selection. *Applied Intelligence, 48*(10), 3462–3481. https://doi.org/10.1007/s10489-018-1158-6

Seifi, A., Ehteram, M., & Soroush, F. (2020). Uncertainties of instantaneous influent flow predictions by intelligence models hybridized with multi-objective shark smell optimization algorithm. *Journal of Hydrology, 587*, 124977. https://doi.org/10.1016/j.jhydrol.2020.124977

Seifi, A., Ehteram, M., & Dehghani, M. (2021). A robust integrated Bayesian multi-model uncertainty estimation framework (IBMUEF) for quantifying the uncertainty of hybrid meta-heuristic in global horizontal irradiation predictions. *Energy Conversion and Management, 241*, 114292.

Singh, N., Singh, S. B., & Houssein, E. H. (2022). Hybridizing salp swarm algorithm with particle swarm optimization algorithm for recent optimization functions. *Evolutionary Intelligence, 15*(1), 23–56. https://doi.org/10.1007/s12065-020-00486-6

Tikhamarine, Y., Souag-Gamane, D., Ahmed, A. N., Kisi, O., & El-Shafie, A. (2020). Improving artificial intelligence models accuracy for monthly streamflow forecasting using grey Wolf optimization (GWO) algorithm. *Journal of Hydrology, 582*, 124435.

Tubishat, M., Ja'afar, S., Alswaitti, M., Mirjalili, S., Idris, N., Ismail, M. A., & Omar, M. S. (2021). Dynamic salp swarm algorithm for feature selection. *Expert Systems with Applications, 164*, 113873. https://doi.org/10.1016/j.eswa.2020.113873

Tubishat, M., Ja'afar, S., Idris, N., Al-Betar, M. A., Alswaitti, M., Jarrah, H., Ismal, M. A., & Omar, M. S. (2022). Improved sine cosine algorithm with simulated annealing and singer chaotic map for Hadith classification. *Neural Computing and Applications, 34*(2), 1385–1406.

Ullah, I., Hussain, I., Uthansakul, P., Riaz, M., Khan, M. N., & Lloret, J. (2020). Exploiting multiverse optimization and sine-cosine algorithms for energy management in smart cities. *Applied Sciences (switzerland), 10*(6), 2095. https://doi.org/10.3390/app10062095

Valikhan-Anaraki, M., Mousavi, S. F., Farzin, S., Karami, H., Ehteram, M., Kisi, O., & El-Shafie, A. (2019). Development of a novel hybrid optimization algorithm for minimizing irrigation deficiencies. *Sustainability, 11*(8), 2337.

Yaghi, R. I., Faris, H., Aljarah, I., Al-Zoubi, A. M., Heidari, A. A., & Mirjalili, S. (2020). *Link prediction using evolutionary neural network models* (pp. 85–111). Springer, Singapore. https://doi.org/10.1007/978-981-32-9990-0_6

Chapter 5
Predicting Aluminum Oxide Grade

Abstract This chapter explains the application of Bayesian model averaging and individual models for estimating aluminum oxide grade. The latitude, longitude, and depth of boreholes are used for estimating aluminum oxide grade. In the first level, the multi-layer perceptron (MLP)–particle swarm optimization (PSO), MLP–salp swarm algorithm (SSA), MLP–naked mole rate algorithm (NMR), MLP–genetic algorithm (MLP-GA), and MLP are used for estimating aluminum oxide grade. In the next level, the outputs of these models are inserted into Bayesian model averaging (BMA). The mean absolute errors (MAE) of the BMA, MLP-NMR, MLP-SSA, MLP-PSO, MLP-GA, and MLP models were 10.12, 10.98, 11.12, 12.23, 14.45, and 15.56 at the testing level. The Nash Sutcliffe efficiencies (NSE) of the BMA, MLP-NMR, MLP-SSA, MLP-PSO, MLP-GA, and MLP models were 0.94, 0.92, 0.89, 0.86, 0.84, and 0.82 at the testing level.

Keywords Multilayer perceptron (MLP) · Particle swarm optimization (PSO) · Salp swarm algorithm (SSA) · Estimating ore grade

5.1 Introduction

In the previous chapters, we examined the structure of individual models. An individual model consists of one model, such as the multi-layer perceptron or radial basis function neural network. An individual model is not superior to all models. Studies have shown that ensemble models improve individual models' accuracy. The accuracy of results may be improved by incorporating multiple individual models into an ensemble structure (Gharekhani et al., 2022). Bayesian model averaging (BMA) is one of the most important ensemble models (Darbandsari & Coulibaly, 2020; Liu & Merwade, 2018). The BMA combines the predictions of each base model and determines the final prediction for the unseen data (Doubleday et al., 2021; Seifi et al., 2021). The BMA is a unique tool for addressing model uncertainty (Liu & Merwade, 2018). Using BMA reduces generalization errors in predictions. BMA has been used in different fields. Liu and Merwade (2018) developed a deterministic water stage prediction model using ensemble predictions from different hydraulic models. For predicting wind speed, Wang et al. (2020) proposed the BMA. The BMA was trained

M. Ehteram et al., *Estimating Ore Grade Using Evolutionary Machine Learning Models*, https://doi.org/10.1007/978-981-19-8106-7_5

based on the outputs produced by three base learners. Based on the results, the BMA method improved model accuracy. Using BMA, Gharekhani et al. (2022) assessed the groundwater vulnerability in a study area. The BMA provided lower uncertainties than other models.

5.1.1 Structure of Bayesian Model Averaging

A BMA is a statistical method that combines multiple competing models' inferences and forecasts. By integrating the model products, BMA produces a probability density function. The forecast PDF, p(d), is given by Eq. (5.1) according to the law of total probability (Ehteram et al., 2021):

$$p\left(o|O^{M_1}, O^{M_2}, \ldots, O\right) = \sum_{i=1}^{N} p\left(o|O^{M_i}, O\right) \times p\left(O^{M_i}|O\right) \tag{5.1}$$

where $p\left(o|O^{M_1}, O^{M_2}, \ldots, O\right)$: the BMA predictive PDF, $p\left(O^{M_i}|O\right)$: the posterior probability of the model's M_i prediction, $O^f = O^{M_1}, O^{M_2}, \ldots, O^{M_N}$: the ensemble of model prediction, o: forecasted variable, O: observation, $\sum_{i=1}^{N} p\left(o|O^{M_i}, O\right)$: the posterior distribution of o given the prediction of model M_i, and O: observed data. The weight of each ensemble member represents its posterior probability. Based on assumptions of BMA, forecasts should not differ from observations. For this reason, a bias-correction method must be applied before BMA implementation (Panahi et al., 2021):

$$f^{M_i} = a_i + bO^{M_i} \tag{5.2}$$

where f_i^t: the bias-corrected forecasts, a_i and b: coefficients. Consequently, Eq. (5.1) can be rewritten as follows:

$$p\left(o|O^{M_1}, O^{M_2}, \ldots, O\right) = \sum_{k=1}^{K} \omega_i \times p\left(o|f^{M_i}, O\right) \tag{5.3}$$

where ω_i: weight of each model.

In the original BMA method, the posterior probability follows a normal distribution. By applying Eq. (5.3) and considering the Gaussian distribution, we can calculate the BMA predictive mean and its variance. For estimating the variance and weight of each competing model, the log-likelihood function is more convenient than the likelihood function:

$$l(\theta) = \log\left(\sum_{i=1}^{l} w_i . p\left(o|f^{M_i}, O\right)\right) \tag{5.4}$$

where $l(\theta)$: likelihood function, $\theta = \{(\omega_i, \sigma_i^2), i = 1, 2, .., N\}$and σ_i^2: variance. Raftery et al. (2005) recommended the expectation–maximization (EMA) algorithm to find θ since it is impossible to calculate θ analytically. The latent variable (Jk) was introduced into the optimization procedure to formulate this optimization problem. The EMA includes two levels, the expectation step and the maximization step. In the expectation step, latent variables are computed based on the current values of parameters. The maximization step estimates parameters using the latent variable value calculated in the expectation step. A probabilistic ensemble prediction is generated based on the combination of individual predictions. Generally, Monte Carlo sampling is used for this purpose (Achite et al., 2022):

(1) The probability distribution $[w1,\ w2,\ ...,\ wk]$ can be used to provide an integer value of i from [1, 2, …, k].
(2) Generate a value of o from the PDFs of $g(o|f^{M_i}, \sigma_i^2)$ for BMA.
(3) Steps 1 and 2 should be repeated K times. This study uses 1000 Monte Carlo simulations (K).
(4) Samples should be ranked ascendingly; the 90% uncertainty interval can be calculated between 5 and 95% quantiles.
(5) Two indices were used to quantify the uncertainty of models:

$$\mathrm{RB} = \frac{\sum_{t=1}^{n} (\mathrm{OG}_U - \mathrm{OR}_L)}{N\sigma_o} \tag{5.5}$$

$$P = \frac{\mathrm{NO}}{N} \times 100 \tag{5.6}$$

where OG_U and OG_L: upper and lower bounds of ore grade, NO: the number of observations enveloped by 95% prediction uncertainty (95PPU) band, RB: width of bound, P: percentage of the observed data that falls within the 95% prediction interval, and N: number of data. Models with the highest P and the lowest RB values have the best performances.

5.2 Case Study

North Khorasan has the largest bauxite deposits in Iran, which cover 8 km. There are many faults in the area, which cause the ore to be broken up into several blocks. The orebody in this block has been explored by drilling 72 exploratory boreholes. The site topography and drilling pattern are shown in Fig. 5.1a. A survey of drill holes reveals constant profiles of hard bauxite, clayey soft bauxite, and kaolinite clay, from top to bottom, respectively. Alumina production depends on hard bauxite, so this area of lithology is known as an ore deposit. Three hundred and sixty two samples from boreholes were used in this study to estimate the Al_2O_3 grade in hard bauxite. The histogram of the Al_2O_3 grade can be seen in Fig. 5.1b. Table 5.1 reports the descriptive statistics of Al_2O_3 grade using these samples. Statistical characteristics of Al_2O_3 grade are presented in Table 5.1. The latitude, longitude, and depth of boreholes are used for estimating aluminum oxide grade. In the first level, the multi-layer perceptron (MLP)–particle swarm optimization (PSO), MLP–salp swarm algorithm (SSA), MLP–naked mole rate algorithm (NMR), MLP–genetic algorithm (MLP-GA), and MLP are used for estimating aluminum oxide grade. In the next level, the outputs of these models are inserted into Bayesian model averaging (BMA).

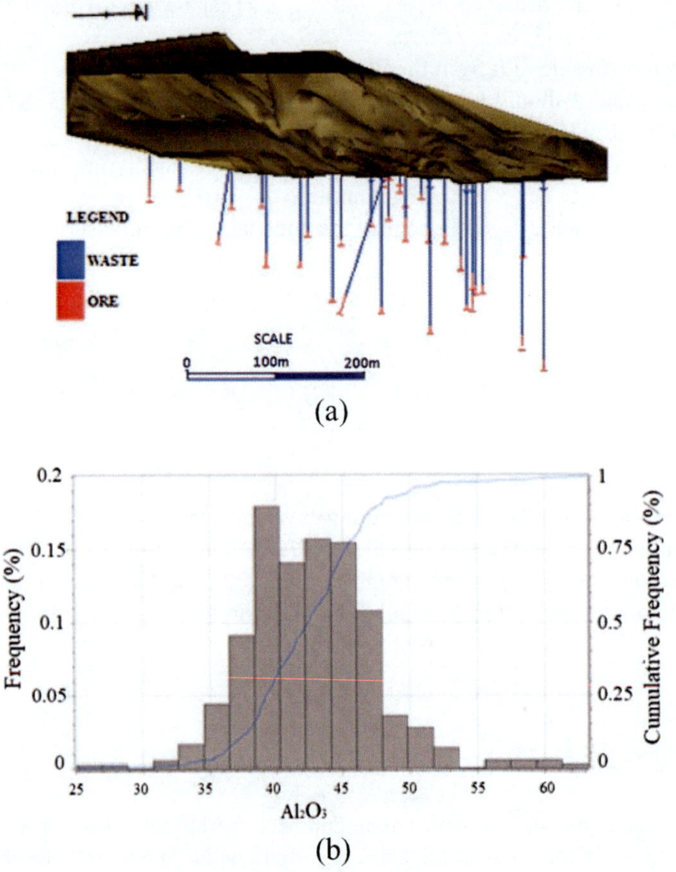

Fig. 5.1 **a** Topography condition and drilled exploratory drill holes in Jajarm deposit, **b** histogram of grade

Table 5.1 Statistical parameters of Al_2O_3

Median	41.62
Mean	42.53
Min	25
Max	63.15
Standard deviation	4.73
Variance	22.46
CV	0.11
Skewness	0.62
Kurtosis	2.49

The error functions are utilized to determine the best model:

1. Root mean square error

$$\mathrm{RMSE} = \sqrt{\frac{\sum_{i=1}^{N} (\mathrm{OG}_{\mathrm{me}} - \mathrm{OG}_{\mathrm{es}})^2}{N}} \tag{5.7}$$

2. Mean absolute error (MAE)

$$\mathrm{MAE} = \frac{1}{N} \sum_{i=1}^{N} |\mathrm{OG}_{\mathrm{me}} - \mathrm{OG}_{\mathrm{es}}| \tag{5.8}$$

3. Percentage of bias (PBIAS)

$$\mathrm{PBIAS} = \frac{\sum_{i=1}^{N} (\mathrm{OG}_{\mathrm{es}} - \mathrm{OG}_{\mathrm{me}})}{\sum_{i=1}^{N} \mathrm{OG}_{\mathrm{me}}} \tag{5.9}$$

4. Nash–Sutcliffe efficiency

$$\mathrm{NSE} = 1 - \frac{\sum_{i=1}^{N} (\mathrm{OG}_{\mathrm{es}} - \mathrm{OG}_{\mathrm{me}})^2}{\sum_{i=1}^{N} \left(\mathrm{OG}_{\mathrm{me}} - \overline{\mathrm{OG}}_{\mathrm{me}}\right)^2} \tag{5.10}$$

5. Centered root mean square difference error (CRMSD)

$$\mathrm{CRMSD} = \sqrt{\frac{\sum_{i=1}^{N} \left(\mathrm{OG}_{\mathrm{me}} - \overline{\mathrm{OG}}_{\mathrm{me}}\right) - \left(\mathrm{OG}_{\mathrm{es}} - \overline{\mathrm{OG}}_{\mathrm{es}}\right)}{N}} \quad (5.11)$$

where $\mathrm{OG}_{\mathrm{me}}$: measured ore grade, $\mathrm{OG}_{\mathrm{es}}$: estimated ore grade, N: number of data, $\overline{\mathrm{OG}}_{\mathrm{me}}$: average measured ore grade, N: number of data, and $\overline{\mathrm{OG}}_{\mathrm{es}}$: average estimated ore grade.

5.3 Results

5.3.1 *Determination of Values of Random Parameters*

The accuracy of optimization algorithms is affected by random parameters. A sensitivity analysis can determine the value of random parameters of the optimization algorithms. In a sensitivity analysis, the objective function values are calculated as a function of random parameters. The best values of random parameters lead to the lowest objective function values. There are two important random parameters: the population size and the maximum number of iterations. The sensitivity analysis in Fig. 5.2a and b determines the number of iterations and the population size. NMR's objective function value ranges from 12.2 to 18.7 for population sizes (Pop) of 100 to 500, as shown in Fig. 5.6a. For NMR, Pop = 200 yielded the lowest objective function. For SSA, PSO, and GA, Pop = 300, Pop = 400, and Pop = 400 provided the lowest objective functions. NMR's objective function value ranges from 12.45 to 18.8 for the maximum number of iterations (MNI) of 50 to 200 (Fig. 5.2b). For NMR, MNI = 100 yielded the lowest objective function. For NMR, SSA, PSO, and GA, the best values of MNI are 100, 150, 200, and 200.

5.3.2 *Investigation of the Accuracy of Models for Predicting Ore Grade*

The MAE, NSE, and PBIAS values of each model are observed in Fig. 5.3a, b, and c. The BMA model reduced the MAE of the MLP-NMR, MLP-SSA, MLP-PSO, MLP-GA, and MLP models by 6.05, 9.8, 18, 25, and 36%, respectively, at the training level. At the training level, the NSE of the BMA model was 0.98, while the NSEs of the MLP-NMR, MLP-SSA, MLP-PSO, MLP-GA, and MLP models were 0.95, 0.92, 0.90, 0.88, and 0.87, respectively. The BMA and MLP provided the lowest and highest PBIAS at the training level. The MAEs of the BMA, MLP-NMR, MLP-SSA, MLP-PSO, MLP-GA, and MLP models were 10.12, 10.98, 11.12, 12.23, 14.45, and 15.56, respectively, at the testing level. The NSEs of the BMA, MLP-NMR, MLP-SSA, MLP-PSO, MLP-GA, and MLP models were 0.94, 0.92, 0.89, 0.86, 0.84, and

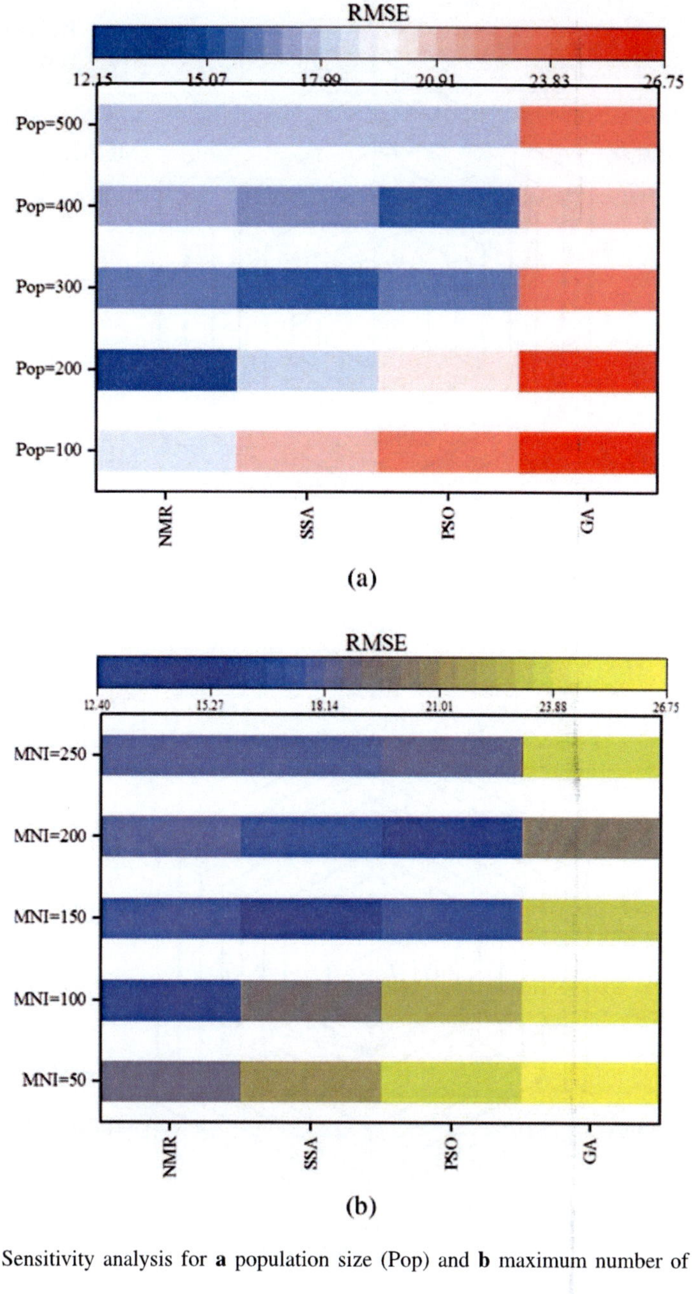

Fig. 5.2 Sensitivity analysis for **a** population size (Pop) and **b** maximum number of iterations (MNOI)

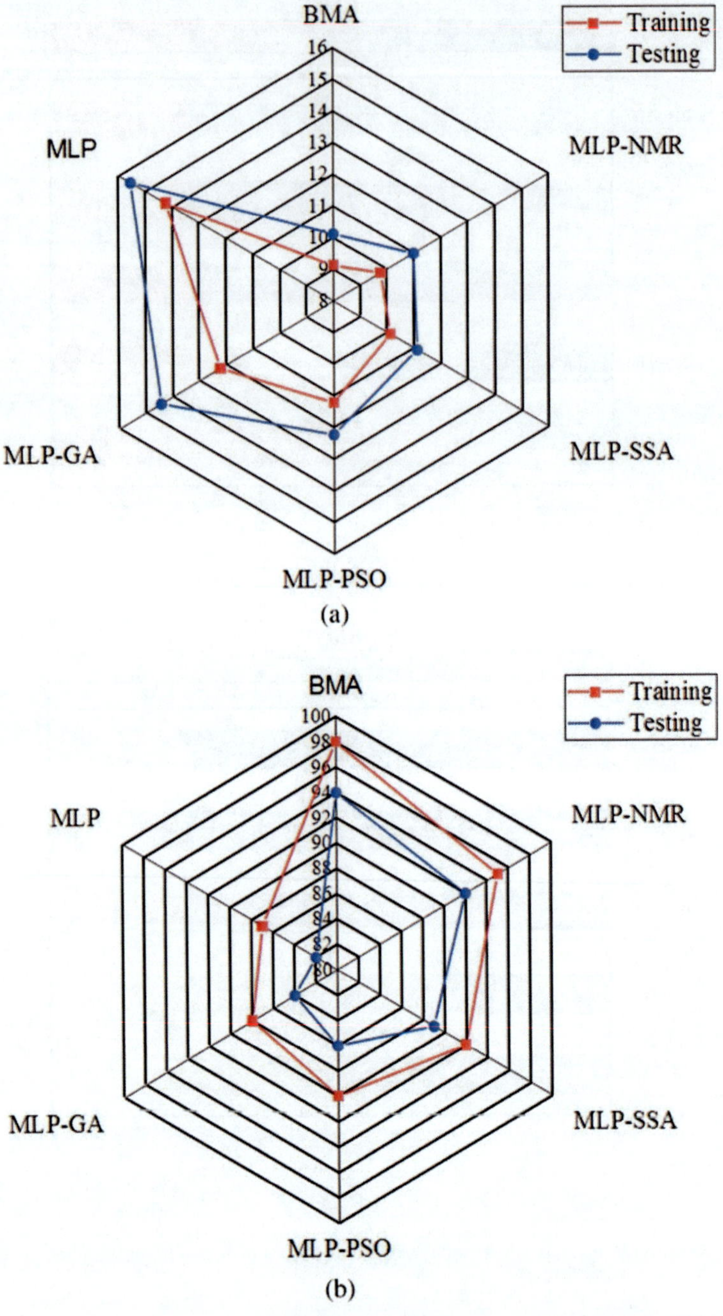

Fig. 5.3 Radar plots for investigating the accuracy of models based on **a** MAE, **b** NSE, and **c** PBIAS

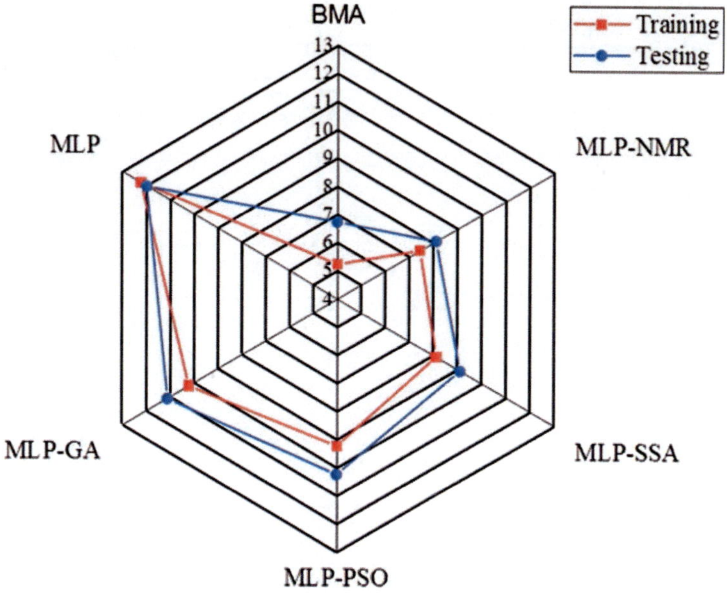

(c)

Fig. 5.3 (continued)

0.82, respectively, at the testing level. The BMA and MLP-NMR provided the lowest PBIAS at the testing level.

Figure 5.4 shows heat scatterplots for different models. Heat scatterplots are an alternative to scatterplots when there are many data points. Heat scatterplots are used to determine the number of data within an area and evaluate accuracy models. The R^2 values of BMA, MLP-NMR, MLP-SSA, MLP-PSO, MLP-GA, and MLP model were 0.9923, 0.9854, 0.9712, 0.9695, 0.9512, and 0.9498, respectively. Generally, most of the data points fall within 33 to 52%.

The Taylor diagram is a graphical tool for assessing model accuracy. Taylor diagrams use CRMSE, correlation coefficient, and standard deviation to determine the accuracy of models. More accurate models are closer to the reference point. A Taylor diagram is used to evaluate the accuracy of models in Fig. 5.5. The Taylor diagram shows that BMA and MLP had the highest and lowest accuracies. The CRMSEs of the BMA, MLP-NMR, MLP-SSA, MLP-PSO, MLP-GA, and MLP models were 0.57, 1.01, 1.53, 2.06, 2.28, and 2.94, respectively.

Figure 5.6 shows the boxplots of models. The medians of observed data, BMA, MLP-NMR, MLP-SSA, MLP-PSO, MLP-GA, and MLP model were 39.87, 39.74, 39.64, 39.5, 39.1, 39.00, and 38.69. The minimum of observed data, BMA, MLP-NMR, MLP-SSA, MLP-PSO, MLP-GA, and MLP model were 25, 25, 25, 25, 23, 19, and 17, respectively. The measured data had the best match with BMA and MLP-NMR.

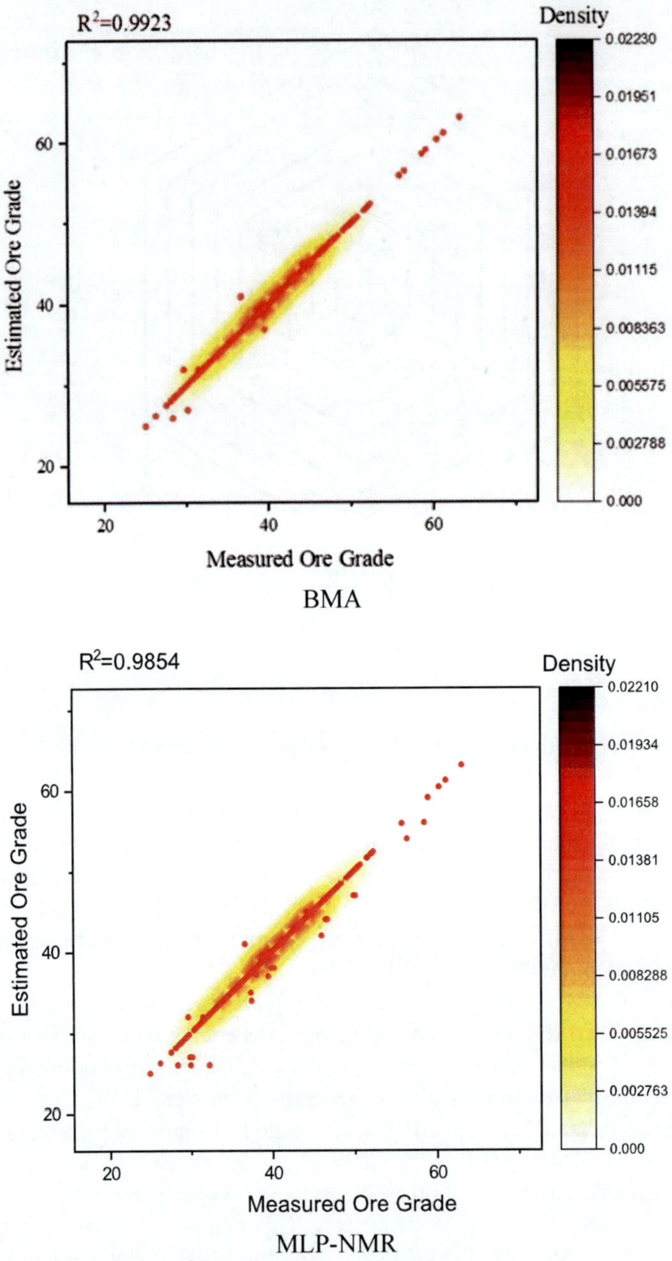

Fig. 5.4 Heat scatterplots of BMA and different MLP models

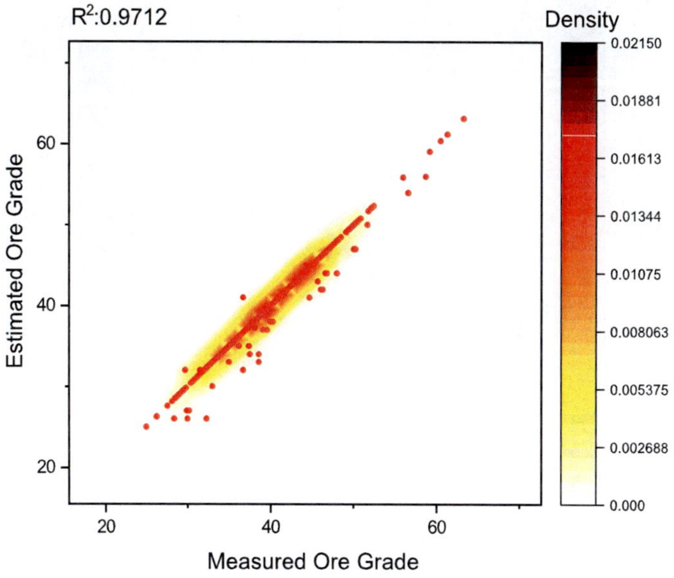

MLP-SSA

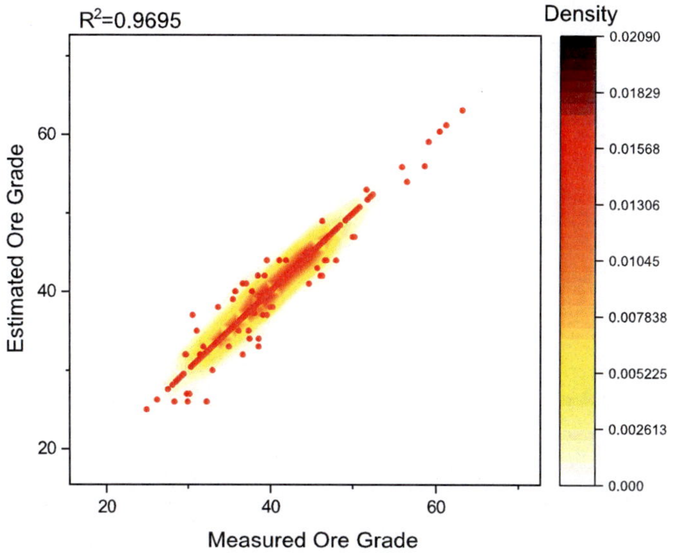

MLP-PSO

Fig. 5.4 (continued)

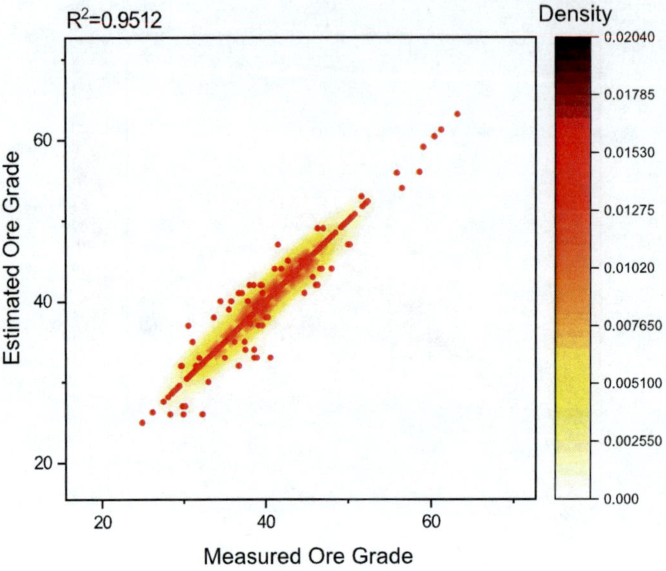

MLP-GA

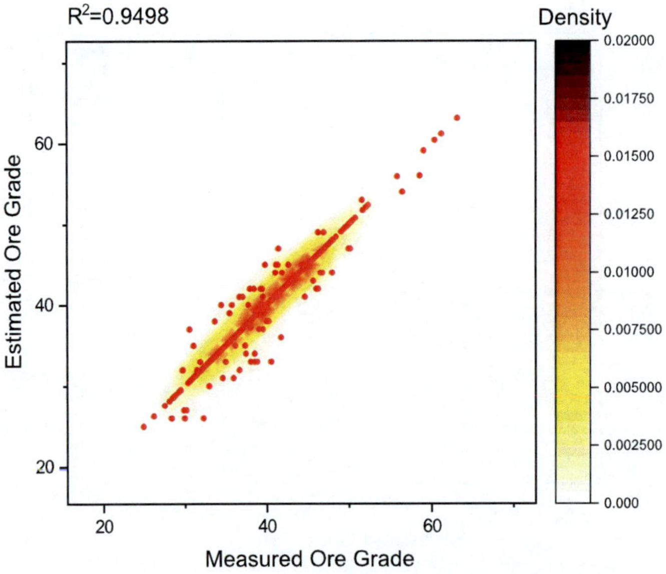

MLP

Fig. 5.4 (continued)

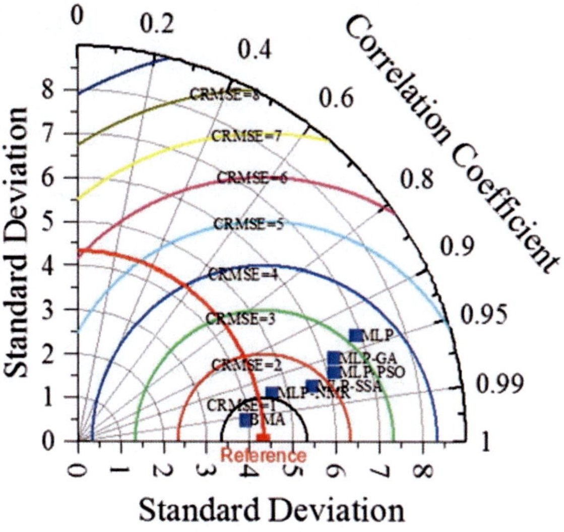

Fig. 5.5 Taylor diagram for comparing models

Figure 5.7 shows a 95% confidence interval for each model at the testing level. The P-values of BMA, MLP-NMR, MLP-SSA, MLP-PSO, MLP-GA, and MLP models were 97, 96, 94, 92, 90, and 88%, respectively. The RB values of BMA, MLP-NMR, MLP-SSA, MLP-PSO, MLP-GA, and MLP models were 12, 15, 18, 20, 22, and 24%, respectively.

Figure 5.8a and b shows the P and RB values at the training level. The RB values of BMA, MLP-NMR, MLP-SSA, MLP-PSO, MLP-GA, and MLP models were 4, 7, 8, 12, 14, and 19%, respectively. The MLP model had the highest uncertainty among the others.

5.3.3 Discussion

Several factors influence the Al_2O_3 ore grade. Thus, mining engineers need robust models to predict Al_2O_3 grade. This study used BMA and optimized MLP models to predict Al_2O_3 ore grade. The optimized MLP models were used individually. BMA used the output of individual models to predict Al_2O_3 ore grade. There are several differences between this study and previous studies:

(1) For predicting ore grade, previous studies used individual soft computing models. No ensemble model was used for predicting Al_2O_3 ore grade. The study uses optimization algorithms to improve the accuracy of individual MLP models. This study also uses ensemble models to predict ore grades.
(2) Due to the high cost and complexity of collecting geological information, the present study only used the geographic coordinates of the boreholes to determine

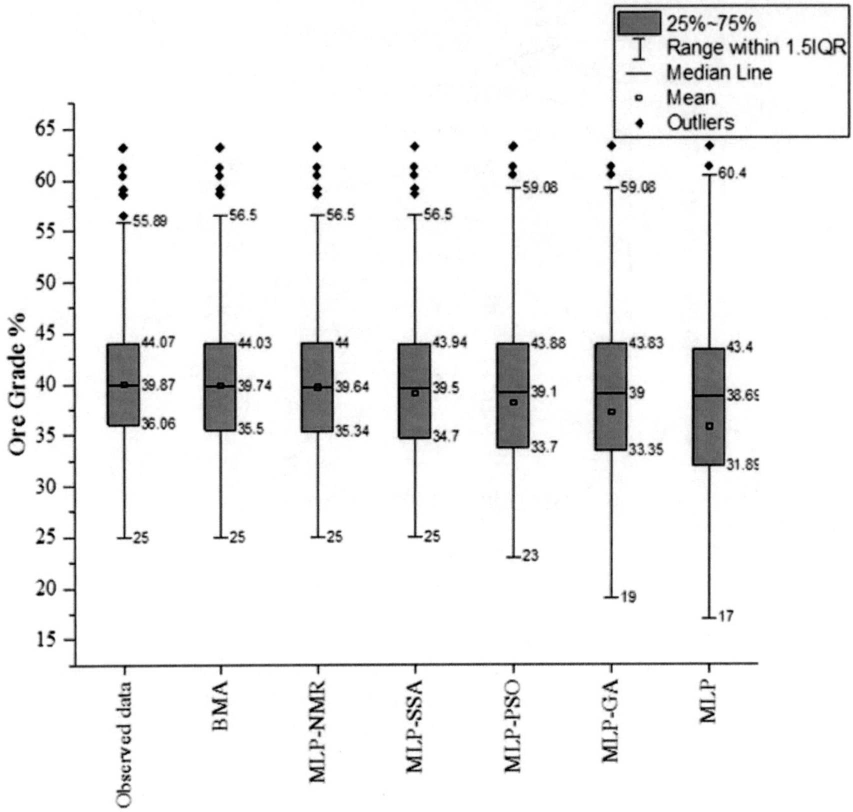

Fig. 5.6 Boxplots of models for comparing models

grade ore. A key advantage of the current study is that MLP models produce accurate results without requiring a lot of input data.

(3) Unlike previous studies, this study examines how uncertainty impacts output. Inputs and parameters are sources of uncertainty. Hence, the uncertainty of parameters and inputs needs to be considered.

In this study, MLP parameters were determined by optimization algorithms. The NMR produced the best results among other optimization algorithms. Breeders strive to improve solutions by following the best breeders. If a breeder cannot improve its fitness, it is placed into the worker group. As a result, local minima are avoided. Because the optimization algorithms use advanced operators, they give different degrees of accuracy. An optimization algorithm can reduce uncertainty parameters by finding the accurate values of model parameters.

The present study results indicated that the BMA gave the best results. However, the modeling process has limitations. BMA assumes that prior distributions follow a normal distribution. This assumption cannot be used to model non-Gaussian forecast

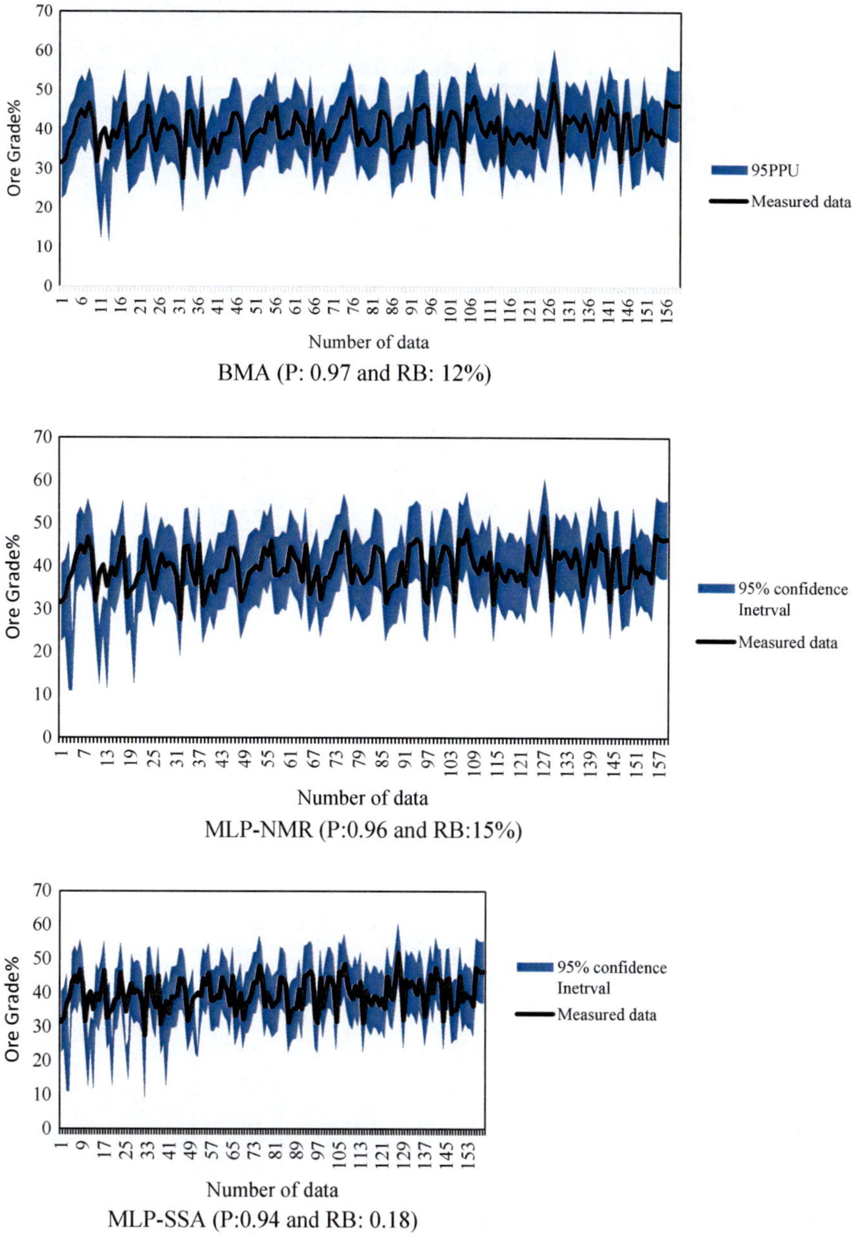

Fig. 5.7 95% confidence interval for the models

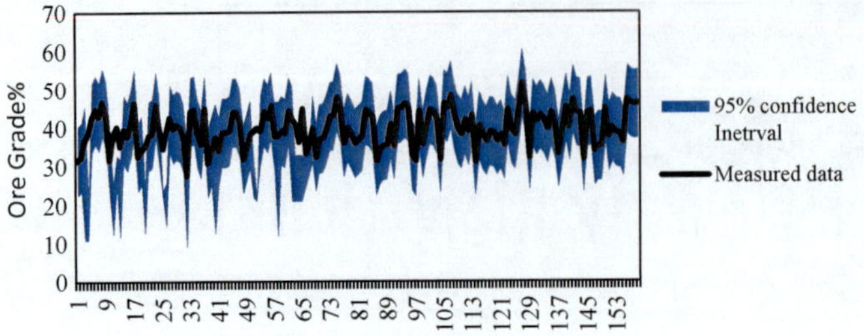

MLP-PSO (P:0.92 and RB:20%)

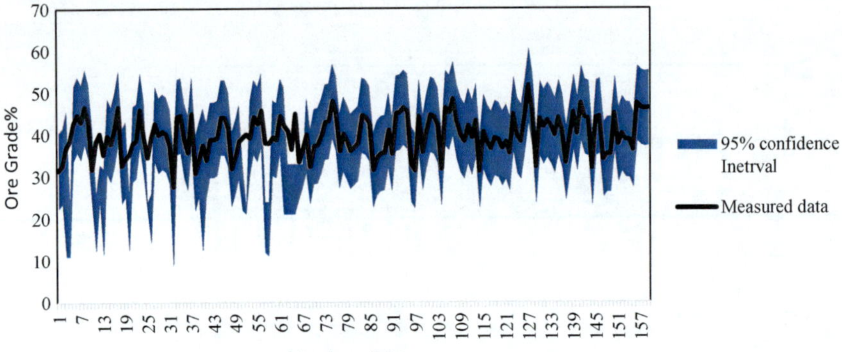

MLP-GA (P:0.90and RB:22%)

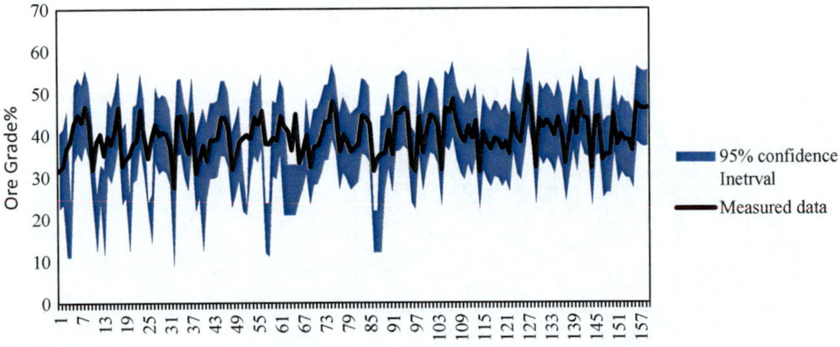

MLP-GA (P:0.88and RB:24%)

Fig. 5.7 (continued)

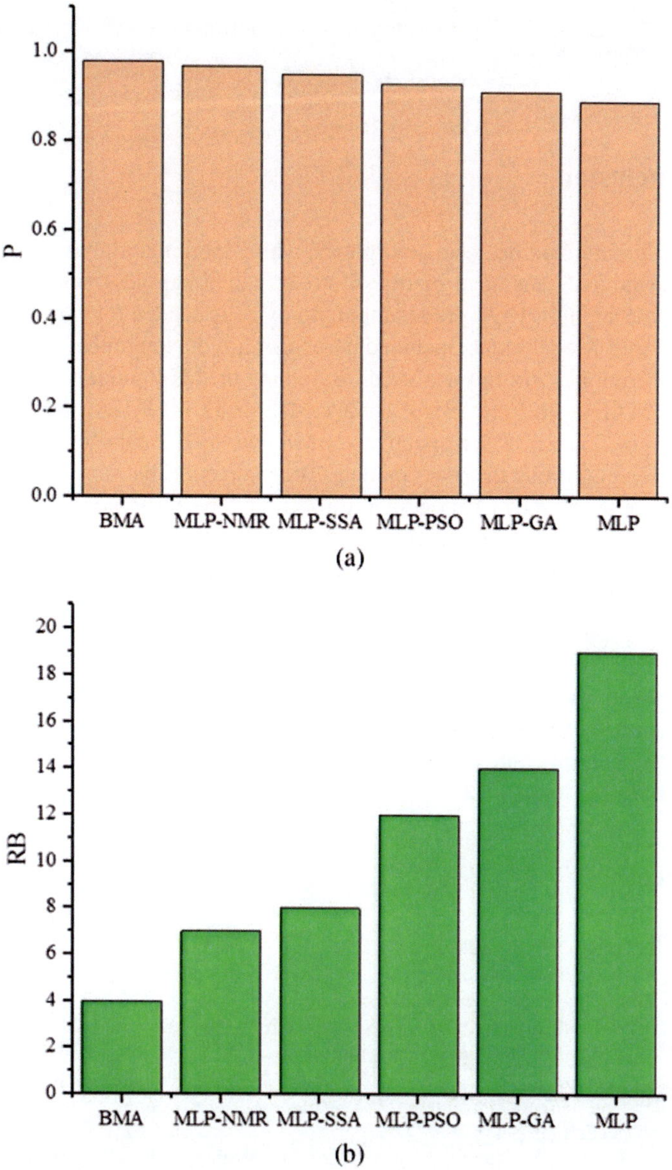

Fig. 5.8 **a** The P-values for the different models and **b** The RB values for the different models at the training level

variables. The next studies can modify this assumption based on the Copula-based Bayesian model.

5.4 Conclusion

Al_2O_3, or aluminum oxide, is an amorphous, white, inert material used in industrial ceramics. Alumina's chemical inertness, strength, and hardness make it a popular basic catalytic support. High-grade aluminum oxide is needed for various manufacturing purposes. Model inputs included borehole latitude, longitude, and depth. The NMR optimization algorithm was used for training the MLP models. BMA reduced the training MAE of MLP-NMR, MLP-SSA, MLP-PSO, MLP-GA, and MLP models by 6.05, 9.8, 18, 25, and 36%, respectively. BMA and MLP had the lowest and highest uncertainty levels among the other models. The main contributions of this paper are the development of the BMA model, optimization of the MLP models, and quantification of the uncertainty of prediction models. The study's main finding was that the BMA model could improve the accuracy of models for predicting ore grade.

Furthermore, the study indicated that the optimized MLP models had lower uncertainty than the standalone MLP models. Mining engineers can use BMAs in the next studies to make accurate grade predictions without incurring high computational costs. We can use other inputs, such as geological information, to develop new estimator models in the next studies. NMR may also be coupled with other soft computing models to predict ore grade in the next research.

References

Achite, M., Banadkooki, F. B., Ehteram, M., Bouharira, A., Ahmed, A. N., & Elshafie, A. (2022). Exploring Bayesian model averaging with multiple ANNs for meteorological drought forecasts. *Stochastic Environmental Research and Risk Assessment*, 1–26.

Darbandsari, P., & Coulibaly, P. (2020). Introducing entropy-based Bayesian model averaging for streamflow forecast. *Journal of Hydrology*. https://doi.org/10.1016/j.jhydrol.2020.125577

Doubleday, K., Jascourt, S., Kleiber, W., & Hodge, B. M. (2021). Probabilistic solar power forecasting using bayesian model averaging. *IEEE Transactions on Sustainable Energy*. https://doi.org/10.1109/TSTE.2020.2993524

Ehteram, M., Ahmed, A. N., Kumar, P., Sherif, M., & El-Shafie, A. (2021). Predicting freshwater production and energy consumption in a seawater greenhouse based on ensemble frameworks using optimized multi-layer perceptron. *Energy Reports, 7*, 6308–6326.

Gharekhani, M., Nadiri, A. A., Khatibi, R., Sadeghfam, S., & Moghaddam, A. A. (2022). A study of uncertainties in groundwater vulnerability modelling using Bayesian model averaging (BMA). *Journal of Environmental Management, 303*, 114168.

Liu, Z., & Merwade, V. (2018). Accounting for model structure, parameter and input forcing uncertainty in flood inundation modeling using Bayesian model averaging. *Journal of Hydrology, 565*, 138–149.

Panahi, F., Ehteram, M., Ahmed, A. N., Huang, Y. F., Mosavi, A., & El-Shafie, A. (2021). Streamflow prediction with large climate indices using several hybrid multilayer perceptrons and copula Bayesian model averaging. *Ecological Indicators, 133*, 108285.
Raftery, A.E., Gneiting, T., Balabdaoui, F., Polakowski, M., 2005. Using Bayesian model averaging to calibrate forecast ensembles. *Monthly weather review, 133*(5), 1155e1174.
Seifi, A., Ehteram, M., & Dehghani, M. (2021). A robust integrated Bayesian multi-model uncertainty estimation framework (IBMUEF) for quantifying the uncertainty of hybrid meta-heuristic in global horizontal irradiation predictions. *Energy Conversion and Management, 241*, 114292.
Wang, G., Jia, R., Liu, J., & Zhang, H. (2020). A hybrid wind power forecasting approach based on Bayesian model averaging and ensemble learning. *Renewable Energy, 145*, 2426–2434.

Chapter 6
Predicting Silicon Dioxide Grade

Abstract This chapter used the IMM and radial basis function neural network models (RBFNN) to predict silicon dioxide grade. First, the RBFNN models were trained using the firefly algorithm (FFA), shark smell optimization (SSO), and sine cosine algorithm (SCA). An inclusive multiple model was built by integrating the outputs of RBFNN-FFA, RBFNN-SSO, RBFNN-SCA, and RBFNN. At the training level, the mean absolute error (MAE) of the IMM was 7.89, while it was 8.12, 8.25, 8.57, and 8.98 for the RBFNN-SSO, RBFNN-SCA, RBFNN-FFA, and RBFNN. At the testing level, the IMM decreased the MAE of the RBFNN-SSO, RBFNN-SCA, RBFNN-FFA, and RFBNN by 0.9, 8.5, 17, and 20%, respectively. The results indicated that the IMM model was reliable for estimating ore grade.

Keywords Optimization algorithms · Inclusive multiple models · Estimating ore grade · Radial basis function neural network

6.1 Introduction

The previous section discussed how ensemble models could improve individual models' accuracy. Bayesian model averaging can enhance the performance of individual models, but it is complex. One of the complexities of the BMA is calculating the prior and posterior distribution. The current chapter presents an ensemble model as an alternative to the BMA. The new ensemble model is called an inclusive multiple model (IMM).

Several artificial neural networks are used to construct the IMM model (Liang et al., 2021). Using the IMM model, predictions are made at two levels. Initially, individual models' outputs were obtained. Next, these outputs are inserted into one type of ANN model (Ehteram et al., 2021). Norouzi et al. (2020) used the IMM model to predict the weir discharge coefficient. Shabani et al. (2021) use the IMM model to predict CO_2 emissions in the agriculture sector. The IMM is recommended as a tool for predicting CO_2 emissions to reduce air pollution.

Khatibi and Nadiri (2021) examined the accuracy of IMMs for predicting groundwater levels. Their study showed that the model was reliable. This chapter used the IMM and radial basis function neural network models (RBFNN) to predict

M. Ehteram et al., *Estimating Ore Grade Using Evolutionary Machine Learning Models*, https://doi.org/10.1007/978-981-19-8106-7_6

silicon dioxide grade. First, the RBFNN models were trained using the firefly algorithm (FFA), shark smell optimization (SSO), and sine cosine algorithm (SCA). An inclusive multiple model was built by integrating the outputs of RBFNN-FFA, RBFNN-SSO, RBFNN-SCA, and RBFNN.

6.1.1 Case Study

The Jajarm bauxite deposit, located 18 km north of Jajarm in North Khorasan, extends more than 8 km. Aluminum oxide is between 30 and 60 percent, and silicon oxide is between 5 and 39 percent in the bauxite layer of this region. Bauxite quality and mineralogy are influenced significantly by tectonic movements. IMM and hybrid RBFNN models were used to predict the silicon oxide grade in bauxite deposits, 72 exploratory boreholes were drilled in this block to determine its geological, lithological, structural, chemical, and structural characteristics. Figure 6.1 shows the topography of this case study. Table 6.1 shows the details of the data. The latitude, longitude, and altitude were used as inputs.

Figure 6.2 shows the data points for the case study. Figure 6.3 shows the structure of IMM model.

The following indices were used to evaluate the models' ability to predict STR in the presence of vegetation cover:

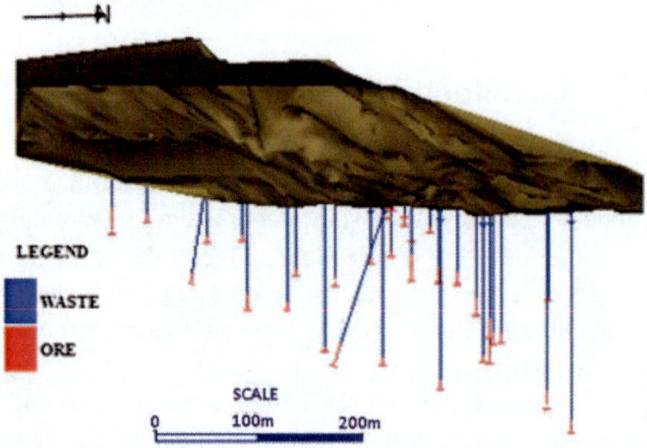

Fig. 6.1 Topography condition in the Jajarm deposit

Table 6.1 The details of the data

Parameter	Value
Average	13.94%
Minimum	5.75%
Maximum	27.24%
Standard deviation	3.82

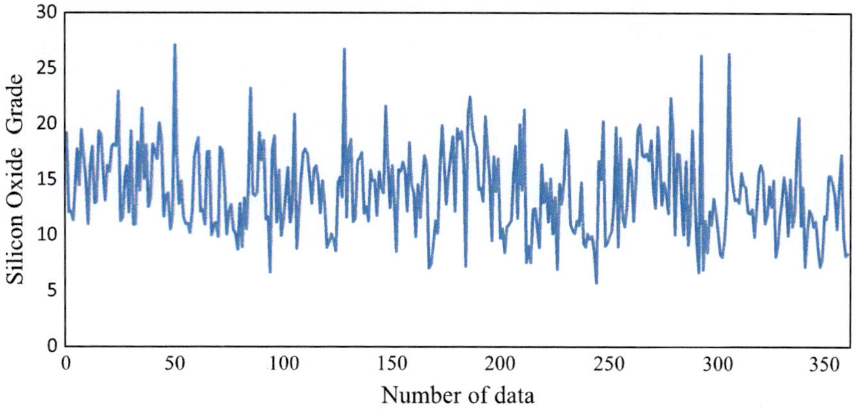

Fig. 6.2 Data points for silicon oxide grade

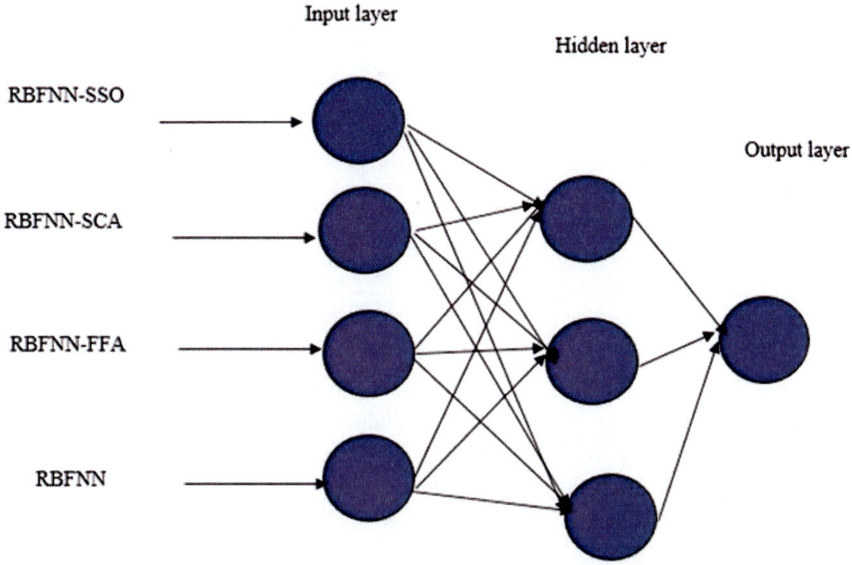

Fig. 6.3 The structure of IMM model

1. Root mean square error

$$\mathrm{RMSE} = \sqrt{\frac{\sum_{i=1}^{N}(\mathrm{OG}_{\mathrm{me}} - \mathrm{OG}_{\mathrm{es}})^2}{N}} \tag{6.1}$$

2. Mean absolute error (MAE)

$$\mathrm{MAE} = \frac{1}{N}\sum_{i=1}^{N} |\mathrm{OG_{me}} - \mathrm{OG_{es}}| \tag{6.2}$$

3. Percentage of bias (PBIAS)

$$\mathrm{PBIAS} = \frac{\sum_{i=1}^{N} (\mathrm{OG_{es}} - \mathrm{OG_{me}})}{\sum_{i=1}^{N} \mathrm{OG_{me}}} \tag{6.3}$$

4. Nash–Sutcliffe efficiency

$$\mathrm{NSE} = 1 - \frac{\sum_{i=1}^{N} (\mathrm{OG_{es}} - \mathrm{OG_{me}})^2}{\sum_{i=1}^{N} \left(\mathrm{OG_{me}} - \overline{\mathrm{OG}}_{\mathrm{me}}\right)^2} \tag{6.4}$$

5. Centered root mean square difference error (CRMSD)

$$\mathrm{CRMSD} = \sqrt{\frac{\sum_{i=1}^{N} \left(\mathrm{OG_{me}} - \overline{\mathrm{OG}}_{\mathrm{me}}\right) - \left(\mathrm{OG_{es}} - \overline{\mathrm{OG}}_{\mathrm{es}}\right)}{N}} \tag{6.5}$$

where $\mathrm{OG_{me}}$: measured ore grade, $\mathrm{OG_{es}}$: estimated ore grade, $\overline{\mathrm{OG}}_{\mathrm{me}}$: average measured ore grade, N: number of data, and $\overline{\mathrm{OG}}_{\mathrm{es}}$: average estimated ore grade. The best model has the lowest RMSE, MAE, PBIAS, and CRMSE values. An ideal model has the highest NSE values.

6.2 Results

6.2.1 Sensitivity Analysis for Choice of Algorithm Parameters

Heat maps of the sensitivity analysis are shown in Fig. 6.4. Random parameters influence the accuracy of optimization algorithms, so determining the parameters' values is important. Figure 6.4a illustrates the variation of population size (POP) versus the objective function value (OBFV). For pop = 150, pop = 300, pop = 450, pop = 600, and pop = 750, the OBFVs of SSO were 12.23, 10.12, 11.45, 11.98, and 12.98, respectively. Therefore, pop = 300 provided the lowest OBFV. Figure 6.4b shows the variation of the maximum number of iterations (MNI) versus the value of the OBF. For MNI = 120, MNI = 240, MNI = 360, MNI = 480, and MNI = 600, the OBFVs of SSO were 12.35, 10.59, 11.78, 11.99, and 12.25. For the FFA and SCA, the POP = 450 provided the lowest OBF. The POP = 450 provided the lowest OBF for the FFA and SCA.

6.2.2 *Investigation of the Accuracy of Models*

The accuracy of the models is investigated in this section. Figure 6.5a shows radar plots based on MAE values. The training MAE of the IMM was 7.89, while it was 8.12, 8.25, 8.57, and 8.98 for the RBFNN-SSO, RBFNN-SCA, RBFNN-FFA, and RBFNN. At the testing level, the IMM decreased the MAE of the RBFNN-SSO, RBFNN-SCA, RBFNN-FFA, and RFBNN by 0.9, 8.5, 17, and 20%, respectively. Figure 6.5b shows radar plots based on NSE values. The training NSE of the IMM was 0.98, while it was 0.95, 0.92, 0.90, and 0.89 for the RBFNN-SSO, RBFNN-SCA, RBFNN-FFA, and RBFNN. The testing NSE of the IMM was 0.97, while it was 0.94, 0.91, 0.89, and 0.82 for the RBFNN-SSO, RBFNN-SCA, RBFNN-FFA, and RBFNN. Figure 6.5c shows radar plots based on PBIAS values. The training NSE of the IMM was 8, while it was 12, 14, 16, and 17 for the RBFNN-SSO, RBFNN-SCA, RBFNN-FFA, and RBFNN. The training and testing values of PBIAS of IMM were 8 and 9%, respectively.

The Taylor diagram is a graphical tool for evaluating the accuracy of models. Taylor diagrams are based on standard deviation, CRMSD, and correlation coefficient. A lower distance from the reference point (measurement value) model is better. Figure 6.6 shows the Taylor diagram. The correlation coefficients of IMM, RBFNN-SSO, RBFNN-SCA, RBFNN-FFA, and RBFNN-FFA were 0.99, 0.95, 0.94, 0.92, and 0.86, respectively. The CRMSD of the IMM, RBFNN-SSO, RBFNN-SCA, RBFNN-FFA, and RBFNN were 1.35, 3.38, 4.06, 4.45, and 5.08, respectively.

Figure 6.7 shows the boxplots of models. The medians of measured data, IMM, RBFNN-SSO, RBFNN-SCA, RBFNN-FFA, and RBFNN-FFA were 18, 19, 21, 23, 26, and 29, respectively. The maximum values of measured data, IMM, RBFNN-SSO, RBFNN-SCA, RBFNN-FFA, and RBFNN-FFA were 38, 39, 40, 42, 44, and 46, respectively. The minimum values of measured data, IMM, RBFNN-SSO, RBFNN-SCA, RBFNN-FFA, and RBFNN-FFA were 4, 4, 6, 8, 9, and 12, respectively.

6.3 Discussion

In the previous section, the IMM model outperformed the optimized RBFNN models. IMM exploits the benefits of multiple RBFNN models. Results of optimized RBFNN models showed that hybrid models outperformed standalone ones. An optimization algorithm using advanced operators can increase the accuracy of RBFNN models. Weight and bias values were unknown parameters in RBFNN models.

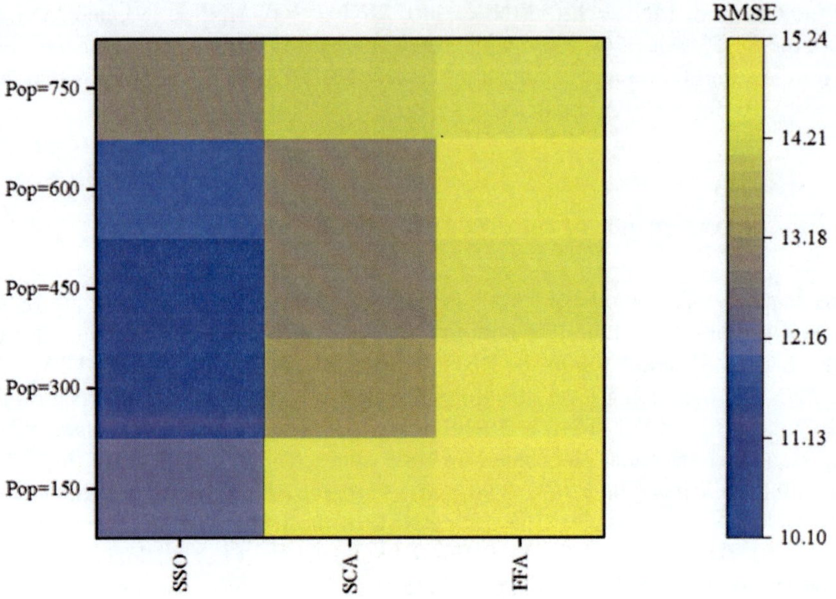

(a)

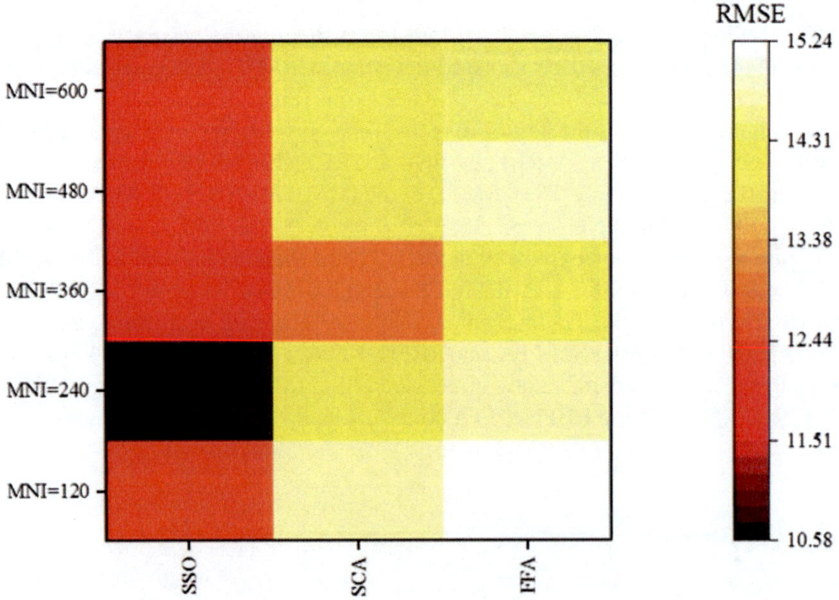

(b)

Fig. 6.4 Sensitivity analysis for **a** Population size (POP) and **b** Maximum number of iterations (MNI)

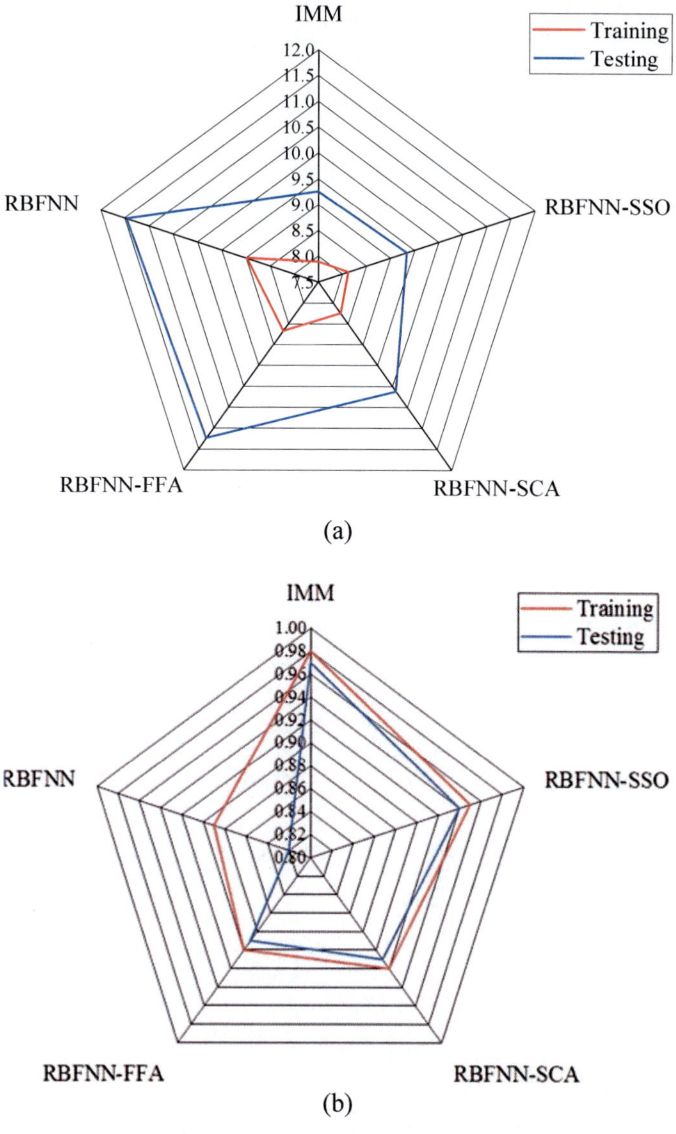

Fig. 6.5 Radar plots **a** MAE, **b** NSE, and **c** PBIAS

An optimization algorithm was used to determine these values. While the multi-layer perceptron models have multiple hidden layers, the RBFNN model only has one. Thus, it lowers unknown parameters. Modeling with IMM is easier than modeling with Bayesian model averaging (BMA). IMM does not require the estimation of prior or posterior distributions.

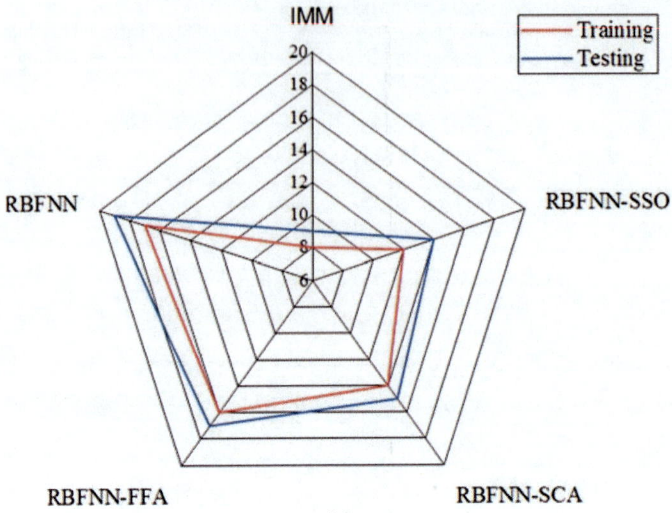

(c)

Fig. 6.5 (continued)

Fig. 6.6 Taylor diagram for the evaluation of the accuracy of models

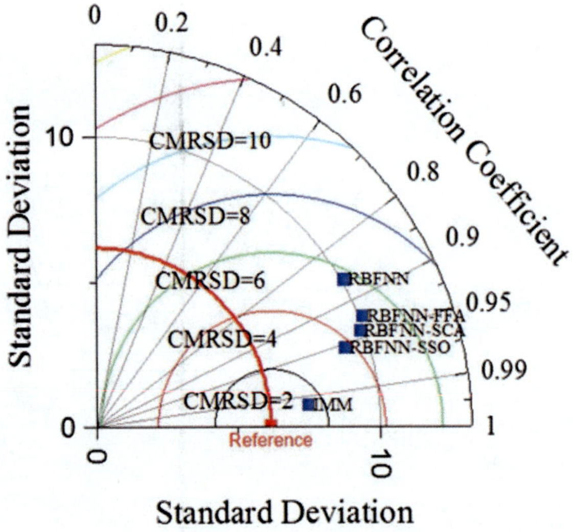

6.4 General Remarks

In this section, the IMM, RBFNN-SSO, RBFNN-SCA, RBFNN-FFA, and RBFNN are used to estimate the grade of silicon dioxide ore. For the RBFNN parameters, weight and bias values were determined using optimization algorithms. IMM models were more accurate than individual RBFNN models. Several RBFNN models are

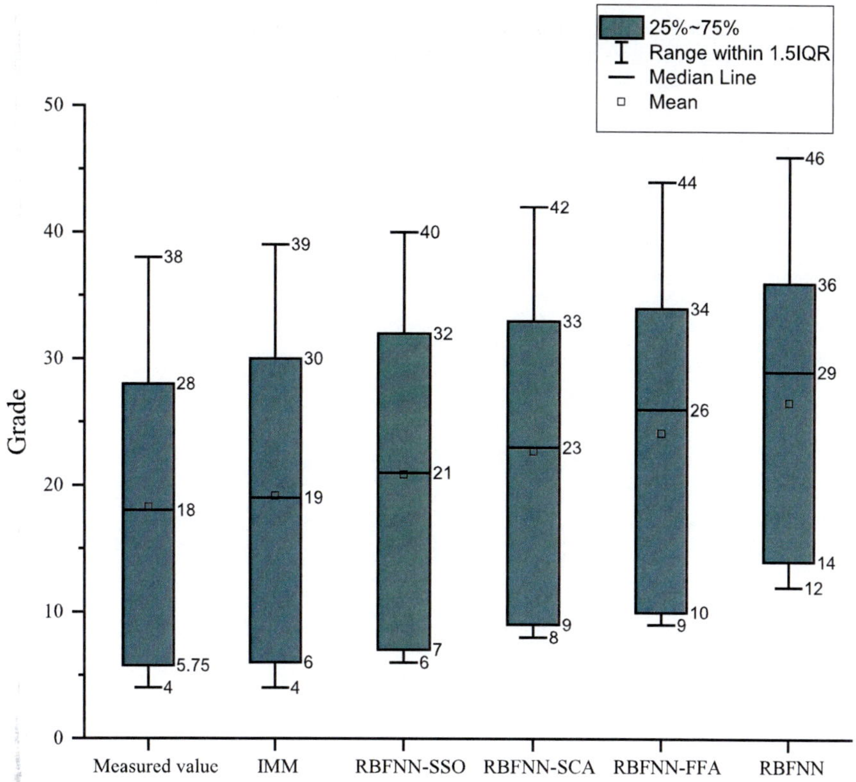

Fig. 6.7 Boxplot for the evaluation of the accuracy of models

incorporated into IMM. Optimizing the RBFNN models could improve their performance. The study results indicated that the IMM model was an excellent alternative to the BMA model. As demonstrated in this chapter, models based on geographic coordinates can accurately predict the grade of silicon dioxide ore. This study can be used in regions where modelers only have geographical data.

References

Ehteram, M., Sammen, S. S., Panahi, F., & Sidek, L. M. (2021). A hybrid novel SVM model for predicting CO2 emissions using multiobjective seagull optimization. *Environmental Science and Pollution Research, 28*(46), 66171–66192.

Khatibi, R., & Nadiri, A. A. (2021). Inclusive multiple models (IMM) for predicting groundwater levels and treating heterogeneity. *Geoscience Frontiers*. https://doi.org/10.1016/j.gsf.2020.07.011

Liang, G., Panahi, F., Ahmed, A. N., Ehteram, M., Band, S. S., & Elshafie, A. (2021). Predicting municipal solid waste using a coupled artificial neural network with archimedes optimisation algorithm and socioeconomic components. *Journal of Cleaner Production, 315*, 128039.

Norouzi, R., Arvanaghi, H., Salmasi, F., Farsadizadeh, D., & Ghorbani, M. A. (2020). A new approach for oblique weir discharge coefficient prediction based on hybrid inclusive multiple model. *Flow Measurement and Instrumentation*. https://doi.org/10.1016/j.flowmeasinst.2020.101810

Shabani, E., Hayati, B., Pishbahar, E., Ghorbani, M. A., & Ghahremanzadeh, M. (2021). A novel approach to predict CO2 emission in the agriculture sector of Iran based on Inclusive Multiple Model. *Journal of Cleaner Production*. https://doi.org/10.1016/j.jclepro.2020.123708

Chapter 7
Predicting Copper Ore Grade

Abstract This chapter estimates copper ore grade using different models. The first part (Part A) investigates the performance of ELM for predicting ore grades. The latitude, longitude, and altitude were used as inputs. The naked mole rat algorithm (NMRA), salp swarm algorithm (SSA), and firefly algorithm (FFA) were used to find the optimal values of input weights of extreme learning machine models (ELM). The second part (Part B) uses the hybrid kriging–ELM for estimating copper ore grade. The latitude, longitude, altitude, and alteration level were used as the inputs. ELM-NMRA decreased the testing mean absolute error (MAE) of the ELM-SSA, ELM-FFA, and ELM models by 14, 17, and 22%, respectively. The testing Nash–Sutcliffe efficiencies (NSE) of the ELM-NMRA, ELM-SSA, ELM-FFA, and ELM were 0.98, 0.94, 0.92, and 0.90. At the testing level, the NSEs of the ELM-NMRA, ELM-SSA, ELM-FFA, and ELM were 0.96, 0.92, 0.90, and 0.89.

Keywords Extreme learning machine models (ELM) models · ELM parameters · Optimization algorithms · Kriging model

7.1 Introduction

Estimating ore grade is a complex and nonlinear process. The previous sections indicated that the optimized artificial neural network models were highly able to calculate ore grade. Furthermore, these models can be employed to improve the performance of classic models such as kriging models. Moreover, these models can be utilized to enhance the performance of traditional models such as kriging models. Kriging is a linear regression technique for estimating ore grade. As the estimation process for ore grade is nonlinear, using Kriging models can provide uncertainty.

Therefore, combining different types of models is necessary to improve prediction accuracy. Hybrid models combine the strengths of multiple component models to improve modeling accuracy and flexibility. Hybrid models combine the strengths of multiple models to improve modeling accuracy and flexibility. Tapoglou et al. (2014) combined the kriging method with an artificial neural network (ANN) to predict groundwater levels. Results indicated that the hybrid method improved the accuracy of the kriging method.

M. Ehteram et al., *Estimating Ore Grade Using Evolutionary Machine Learning Models*, https://doi.org/10.1007/978-981-19-8106-7_7

Seo et al. (2015) used hybrid kriging–artificial neural network models to predict spatial precipitation distributions. The results revealed that the hybrid model provided the best results. Ismail et al. (2019) coupled ANN with kriging to predict mean residence time distribution (MRTD). Based on the results, the model could accurately predict the MRTD.

This chapter combined an extreme learning machine (ELM) model with a kriging model to improve its accuracy for estimating ore grade. A case study used the hybrid models to predict copper ore grade. This section is divided into two parts. The first part (Part A) investigates the performance of ELM for predicting ore grade. The latitude, longitude, and altitude were used as the inputs. The naked mole rat algorithm (NMRA), salp swarm algorithm (SSA), and firefly algorithm (FFA) were used to find the optimal values of input weights of ELM. The second part (Part B) uses the hybrid kriging–ELM for estimating copper ore grade. The latitude, longitude, altitude, and alteration level were used as the inputs. The second part used the NMRA, SSA and FFA to find the optimal values of input weights of ELM.

7.1.1 Kriging Method

An ordinary kriging estimator uses spatial interpolation to estimate data (Ali Akbar, 2012). In this method, the data sets are assumed to be stationary, and the semivariogram is used to estimate values without bias and with the least variance. The Ok is based on the following equation (Ali Akbar, 2012):

$$\hat{Y}(x_0) = \sum_{i=1}^{n} \lambda_i Y(x_i) \tag{7.1}$$

where $\hat{Y}(x_0)$: the estimated values at place x_o, $Y(x_i)$: the sampled value at place x_0, and λ_i: the weighting factor. The estimation error is computed as follows:

$$\hat{Y}(x_0) - Y(x_0) = R(x_0) = \sum_{i=1}^{n} \lambda Y(x_i) - Y(x_0) \tag{7.2}$$

If an estimator is considered unbiased, the mean of the estimated error must be equal to zero.

$$E\{R(x_0)\} = 0 \tag{7.3}$$

Minimizing the estimation error variance leads to linear algebraic equations for the weighting factors. A Lagrange multiplier method and an optimization routine can be used to solve the problem.

7.2 Hybrid ELM and Kriging Method

The hybrid ELM–kriging method has the following levels:

(1) First, ELM models run using training and testing data.
(2) Then, the residual values based on the outputs of ELM and measured values are estimated:

$$\nu = \mathrm{GR}_{\mathrm{me}} - \mathrm{GR}_{\mathrm{es}} \tag{7.4}$$

where ν: residual, $\mathrm{GR}_{\mathrm{me}}$: measured values, and $\mathrm{GR}_{\mathrm{es}}$: estimated values by the ELM.
(3) The kriging method is used as follows:

$$\hat{\nu}(x_i) = \sum_{i=1}^{n} \lambda_i \nu(x_i) \tag{7.5}$$

where $\hat{\nu}(x_i)$: the estimated ν at a location $\mathrm{x_i}$ with the kriging method.
(4) The final is computed as follows:

$$\mathrm{out}(x_i) = \hat{\nu}(x_i) + \mathrm{GR}_{\mathrm{me}} \tag{7.6}$$

where $\mathrm{out}(x_i)$: final output.

7.3 Case Study

This study examined the Iju porphyry copper deposit in the Kerman Province of Iran as one of the important copper deposits. The region is placed at the latitude of 30°31′45″ to 30°33′5′ and the longitude 56°10′54″ to 54′57″30. The copper in the Iju deposit comes mainly from micro-volcanic rocks. The micro-volcanic rocks in Iju have been poorly studied. Hence, the Iju deposits of Kerman need to be fully investigated. Figure 7.1 shows data points. The alteration levels include potassic and phyllic.

Neural network (NN) algorithms perform numerical calculations, which means they can only deal with numeric data. Since the alteration types were inputs to the models, they must be converted to a numerical value. The alteration types were converted into Boolean variables for feeding the ANN.

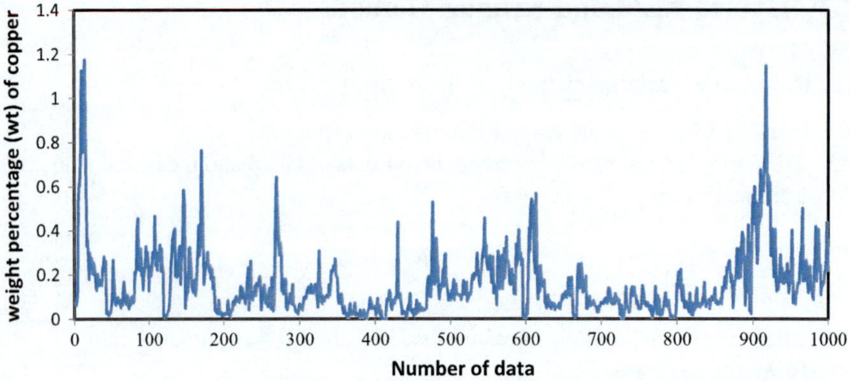

Fig. 7.1 Data points for the case study

The error indices are utilized to assess the potential of models

1. Root mean square error

$$\mathrm{RMSE} = \sqrt{\frac{\sum_{i=1}^{N} (\mathrm{OG}_{\mathrm{me}} - \mathrm{OG}_{\mathrm{es}})^2}{N}} \tag{7.7}$$

2. Mean absolute error (MAE)

$$\mathrm{MAE} = \frac{1}{N} \sum_{i=1}^{N} |\mathrm{OG}_{\mathrm{me}} - \mathrm{OG}_{\mathrm{es}}| \tag{7.8}$$

3. Percentage of bias (PBIAS)

$$\mathrm{PBIAS} = \frac{\sum_{i=1}^{N} (\mathrm{OG}_{\mathrm{es}} - \mathrm{OG}_{\mathrm{me}})}{\sum_{i=1}^{N} \mathrm{OG}_{\mathrm{me}}} \tag{7.9}$$

4. Nash–Sutcliffe efficiency

$$\mathrm{NSE} = 1 - \frac{\sum_{i=1}^{N} (\mathrm{OG}_{\mathrm{es}} - \mathrm{OG}_{\mathrm{me}})^2}{\sum_{i=1}^{N} \left(\mathrm{OG}_{\mathrm{me}} - \overline{\mathrm{OG}}_{\mathrm{me}}\right)^2} \tag{7.10}$$

5. Centered root mean square difference error (CRMSD)

$$\mathrm{CRMSD} = \sqrt{\frac{\sum_{i=1}^{N} \left(\mathrm{OG_{me}} - \overline{\mathrm{OG}}_{\mathrm{me}}\right) - \left(\mathrm{OG_{es}} - \overline{\mathrm{OG}}_{\mathrm{es}}\right)}{N}} \tag{7.11}$$

where $\mathrm{OG_{me}}$: measured ore grade, $\mathrm{OG_{es}}$: estimated ore grade, N: number of data, $\overline{\mathrm{OG}}_{\mathrm{me}}$: average measured ore grade, N: number of data, and $\overline{\mathrm{OG}}_{\mathrm{es}}$: average estimated ore grade. The best model has the lowest RMSE, MAE, PBIAS, and CRMSE values. An ideal model has the highest NSE values.

7.4 Part A

7.4.1 Results of Part A

7.4.1.1 Sensitivity Analysis for Selection of Random Parameters

The first step is to determine the values of random parameters. Figure 7.2a shows how sensitivity analysis is performed for algorithms. For the NMRA, the objective function values (OBFVs) of the population size (Pop) = 50, Pop = 100, Pop = 150, Pop = 200, Pop = 250 were 7.23, 6.56, 7.14, 7.85, and 7.93, respectively. Thus, the Pop = 100 provided the lowest value of the (OBF). For the SSA, the objective function values (OBFVs) of the Pop = 50, Pop = 100, Pop = 150, Pop = 200, Pop = 250 were 8.12, 8.00, 7.99, 8.12, and 8.89, respectively. Thus, the Pop = 100 provided the lowest value of the (OBF). Figure 7.2b shows how sensitivity analysis is performed for algorithms. For the NMRA, the objective function values of the maximum number of iterations (MNI) = 50, MNI = 100, MNI = 150, MNI = 200, MNI = 250 were 7.66, 6.59, 7.32, 7.89, and 7.99, respectively. Thus, the Pop = 100 provided the lowest value of the OBF. For the SSA, the objective functions of MNI = 50, MNI = 100, MNI = 150, MNI = 200, MNI = 250 were 8.42, 8.23, 7.85, 9.19, and 9.2, respectively. The MNI and POP were obtained for the FFA through a similar process.

7.4.1.2 Investigation of Different Accuracies of Models

In this section, the accuracies of models are investigated. Figure 7.3a shows radar plots based on MAE values. The training MAEs of the ELM-NMRA, ELM-SSA, ELM-FFA, and ELM were 5.98, 6.12, 6.99, and 7.23. ELM-NMRA decreased the testing MAE of the ELM-SSA, ELM-FFA, and ELM models by 14, 17, and 22%, respectively. Training NSEs of the ELM-NMRA, ELM-SSA, ELM-FFA, and ELM were 0.98, 0.94, 0.92, and 0.90. At the testing level, the NSEs of the ELM-NMRA,

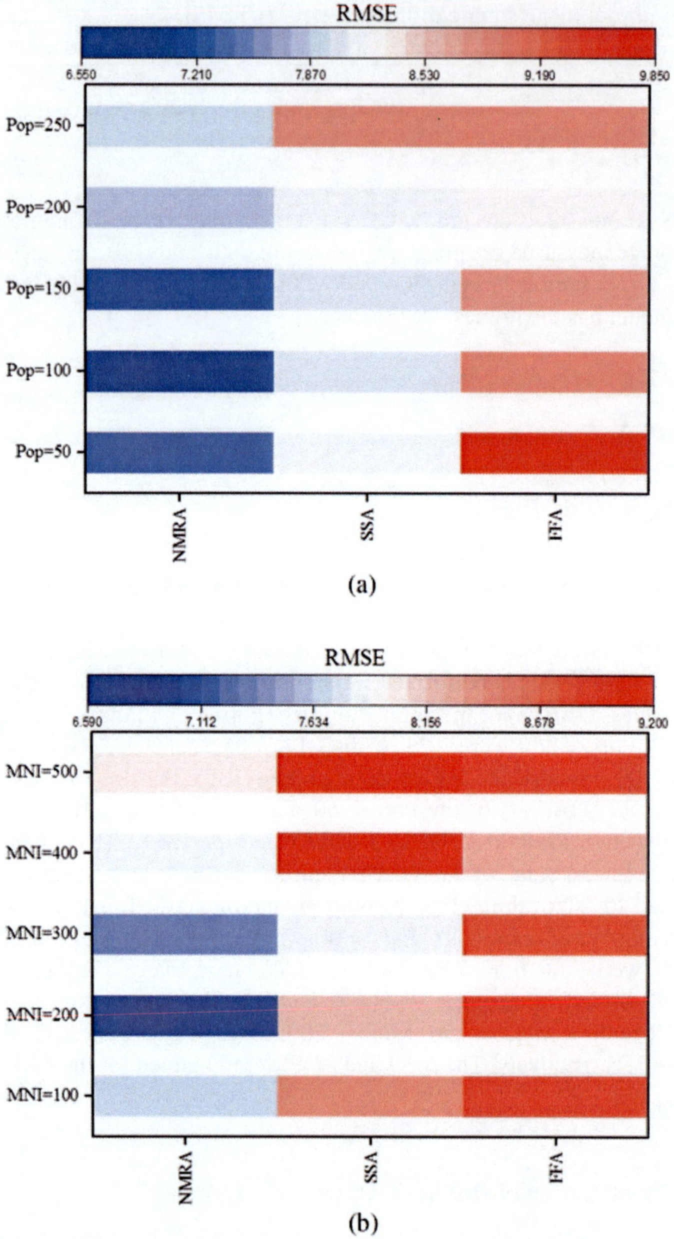

Fig. 7.2 Sensitivity analysis finding **a** PoP and **b** MNI

ELM-SSA, ELM-FFA, and ELM were 0.96, 0.92, 0.90, and 0.89, respectively. The training PBIASs of the ELM-NMRA, ELM-SSA, ELM-FFA, and ELM were 0.12, 0.14, 0.17, and 0.19. The testing PBIASs of the ELM-NMRA, ELM-SSA, ELM-FFA, and ELM were 0.14, 0.16, 0.18, and 0.22.

Taylor diagrams are used to assess the accuracy of models (Fig. 7.4). Using standard deviation, correlation coefficient, and CRMSD, Taylor diagrams evaluate the accuracy of models. The model with the lowest distance from the reference point performs best. The CRMSDs of the ELM-NMRA, ELM-SSA, ELM-FFA, and ELM models were 1.5, 2.21, 3.59, and 5.97, respectively. The correlation coefficients of the ELM-NMRA, ELM-SSA, ELM-FFA, and ELM models were 0.9846, 0.9802, 0.9377, and 0.9288, respectively.

Figure 7.5 shows the boxplots of models for predicting copper grade at the testing level. The medians of measured data, ELM-NMRA, ELM-SSA, ELM-FFA, and ELM models were 0.3642, 0.3671, 0.4014, 0.4185, and 0.495, respectively.

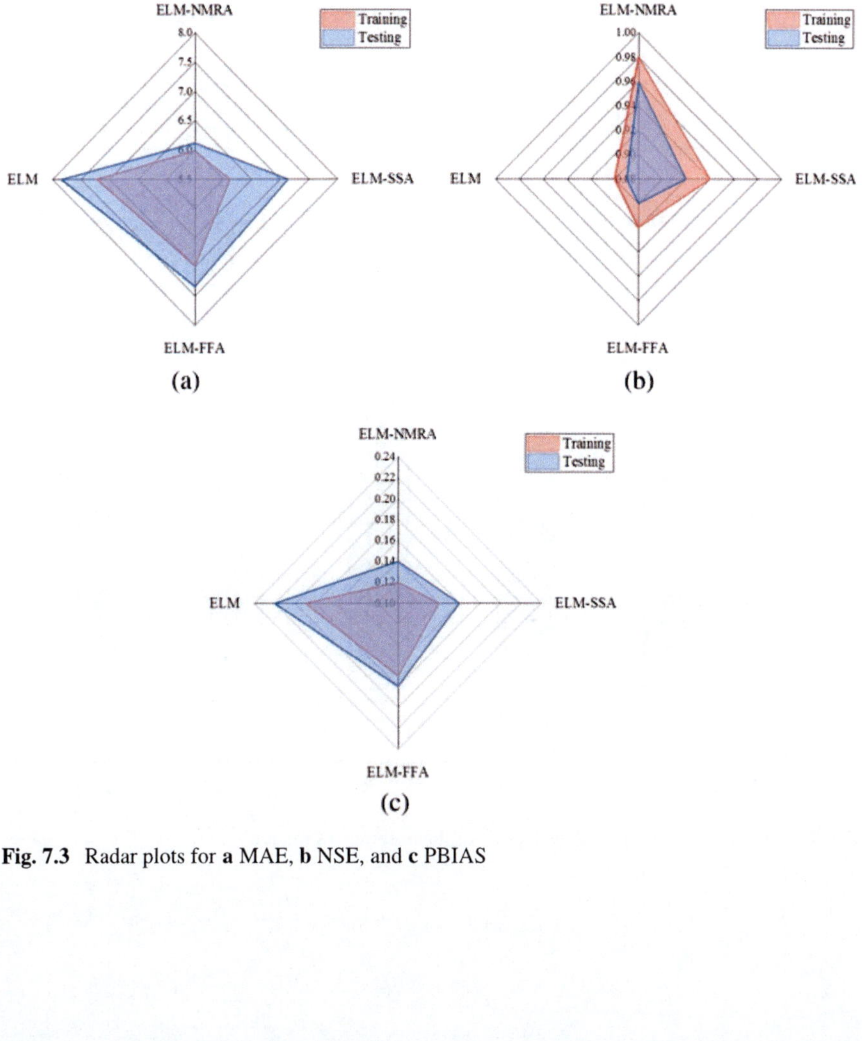

Fig. 7.3 Radar plots for **a** MAE, **b** NSE, and **c** PBIAS

Fig. 7.4 Taylor diagram for evaluation of the accuracy of models

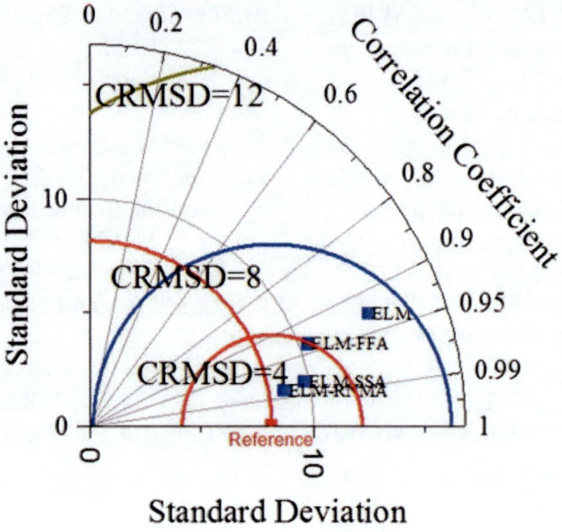

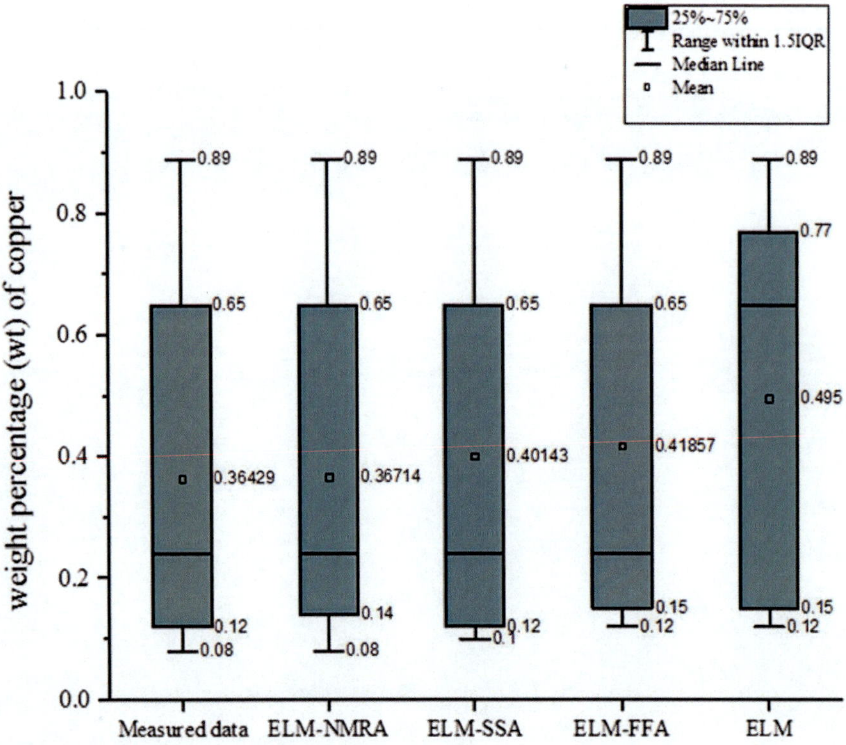

Fig. 7.5 Boxplots of models for evaluation of the accuracy of models

7.5 Part B

7.5.1 Results for Part B

7.5.1.1 Investigation of the Accuracy of Hybrid Kriging–ELM Models

This section examines the performance of hybrid Kriging (KR)–ELM models and standalone ELM and KR models for predicting copper grade. Figure 7.6a evaluates the accuracy of hybrid ELM-KR models based on MAE values. The training MAEs of the ELM-KR-NMRA, ELM-KR-SSA, ELM-KR-FFA, ELM, and KR were 3.25, 4.12, 4.45, 4.55, and 4.89. ELM-NMRA decreased the testing MAE of the ELM-SSA, ELM-FFA, ELM, and KR models by 13, 15, 16, and 20%, respectively. Figure 7.6b evaluates the accuracy of hybrid ELM-KR models based on NSE values. At the training level, the NSEs of the ELM-KR-NMRA, ELM-KR-SSA, ELM-KR-FFA, ELM, and KR models were 0.99, 0.97, 0.95, 0.94, and 0.93. Figure 7.6c evaluates the accuracy of hybrid ELM-KR models based on PBIAS values. At the training level, the PBIASs of the ELM-KR-NMRA, ELM-KR-SSA, ELM-KR-FFA, ELM, and KR models were 8, 11, 12, 14, and 16.

Taylor diagrams for the KR-ELM models are shown in Fig. 7.7. The CRMSDs of the ELM-KR-NMRA, ELM-KR-SSA, ELM-KR-FFA, ELM, and KR models were 0.71, 2.44, 3.22, 5.94, and 6.36, respectively. The correlation coefficients of the ELM-KR-NMRA, ELM-KR-SSA, ELM-KR-FFA, ELM, and KR models were 0.99, 0.98, 0.97, 0.94, and 0.74, respectively.

7.6 General Remarks

This section predicted copper grade using hybrid kriging–ELM models. The results showed that the ELM model outperformed the Kriging method. It was found that the hybrid ELM–kriging method improved the accuracy of the ELM and kriging methods. Hence, coupling kriging with other methods can improve the accuracy of kriging methods. Besides predicting copper grade, models of the current chapter can be applied in other fields. The improved kriging method can also predict other target variables such as groundwater level, evaporation, rainfall, and temperature.

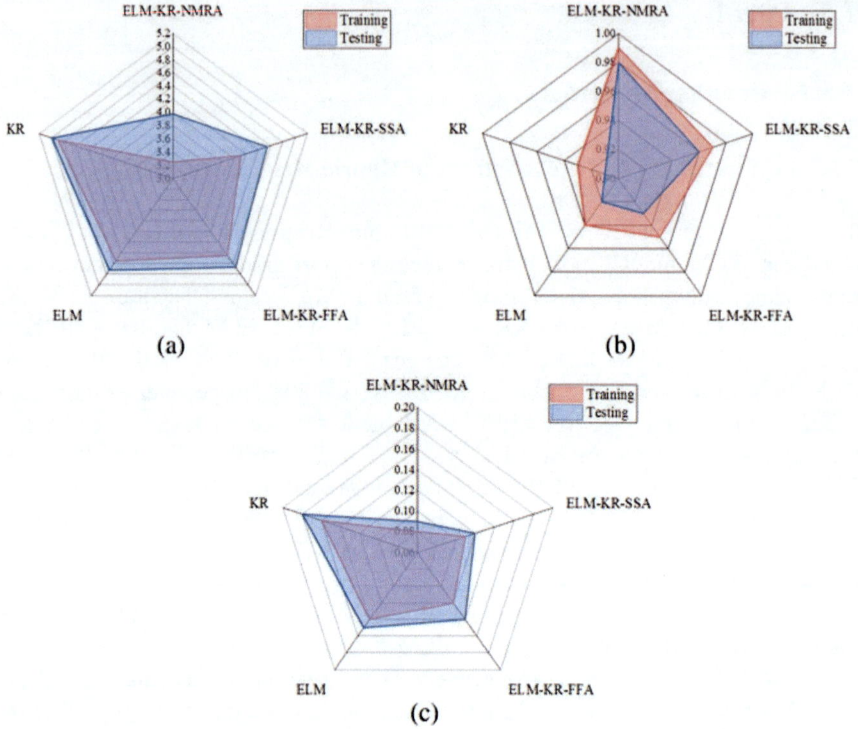

Fig. 7.6 Radar plots for evaluation of the accuracies of ELM-KR models based on **a** MAE, **b** NSE, and **c** PBIAS

Fig. 7.7 Efficiency assessment of models based on Taylor diagram

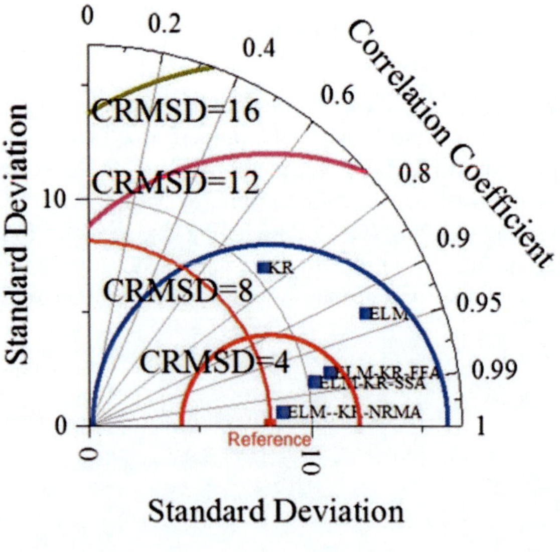

References

Ali Akbar, D. (2012). Reserve estimation of central part of Choghart north anomaly iron ore deposit through ordinary kriging method. *International Journal of Mining Science and Technology*. https://doi.org/10.1016/j.ijmst.2012.01.022

Ismail, H. Y., Singh, M., Darwish, S., Kuhs, M., Shirazian, S., Croker, D. M., Khraisheh, M., Albadarin, A. B., & Walker, G. M. (2019). Developing ANN-Kriging hybrid model based on process parameters for prediction of mean residence time distribution in twin-screw wet granulation. *Powder Technology*. https://doi.org/10.1016/j.powtec.2018.11.060

Seo, Y., Kim, S., & Singh, V. P. (2015). Estimating spatial precipitation using regression kriging and artificial neural network residual kriging (RKNNRK) hybrid approach. *Water Resources Management*. https://doi.org/10.1007/s11269-015-0935-9

Tapoglou, E., Karatzas, G. P., Trichakis, I. C., & Varouchakis, E. A. (2014). A spatio-temporal hybrid neural network-kriging model for groundwater level simulation. *Journal of Hydrology*. https://doi.org/10.1016/j.jhydrol.2014.10.040

Chapter 8
Estimating Iron Ore Grade

Abstract This chapter uses the group machine data handling (GMDH) models for predicting iron ore grade. The sine cosine algorithm (SCA), salp swarm algorithm (SSA), particle swarm optimization (PSO), and firefly algorithm (FFA) were used to optimize the GMDH parameters. Since model parameters and inputs are sources of uncertainty, model uncertainty must be quantified. At the training level, the mean absolute errors (MAEs) of the GMDH-SSA, GMDH-SCA, GMDH-FFA, GMDH-PSO, GMDH-GA, and GMDH were 4.55, 5.12, 5.54, 5.89, and 5.91. At the testing level, the GMDH-SSA decreased the MAE of the GMDH-SCA, GMDH-FFA, GMDH-PSO, GMDH-GA, and GMDH by 4.7, 14, 16, and 17%, respectively. The optimized GMDH models had a high potential for estimating iron ore grade.

Keywords Sine cosine algorithm · Group machine data handling · Iron ore grade estimation · Firefly algorithm

8.1 Introduction

Industry development and infrastructure expansion are necessary for economic growth. For industry development, raw materials are essential. Population growth, energy scarcity, and raw material shortages pose challenges in different countries (Shahbeik et al., 2014). In general, establishing an industrial unit depends on the availability of raw materials. Iron is one of the essential raw materials. Iron has played a crucial role in modern civilization's creation, growth, and development (Bargawa & Tobing, 2020). The largest importer of iron ore is China. Iron ore prices are rising, which can hinder the development of industrial projects. The presence of silica, sulfur, copper, carbonate, alumina, titanium, and phosphorus in iron ore deposits affects the price and quality of the product. Steel's mechanical properties are adversely affected by phosphorus and sulfur. Iron is a significant raw material for railroad construction (Afeni et al., 2020). One of the essential requirements of the industry is high-grade iron ore (Shamsi et al., 2021). Locating iron-rich areas is possible by analyzing iron grades. Finding iron-rich areas will reduce energy consumption during drilling.

M. Ehteram et al., *Estimating Ore Grade Using Evolutionary Machine Learning Models*, https://doi.org/10.1007/978-981-19-8106-7_8

This chapter uses the group machine data handling (GMDH) models for predicting iron ore grade. The sine cosine algorithm (SCA), salp swarm algorithm (SSA), particle swarm optimization (PSO), and firefly algorithm (FFA) were used to optimize the GMDH parameters. The part A of the current chapter investigates the performance of hybrid GMDH models for estimating iron grade. The current chapter examines the uncertainty of models in Part B. Since model parameters and inputs are sources of uncertainty, model uncertainty must be quantified.

8.2 Part A

8.2.1 Case Study

Jalal-Abad iron ore deposit (IOD) is located in Iran's Kerman Province. Jalal-Abad iron ore is mainly composed of sedimentary, volcanic, and volcanic rocks. Jalalabad has one of the largest iron ore reserves in Iran. This mine has a reserve of 187 million tons with an average grade of 43%. There are 1000 data points 80 and 2% of data points were used for training and testing levels. In this study, the latitude, longitude, and depth of boreholes were used as the inputs to estimate iron ore grade. Table 8.1 provides the details of the data. Figure 8.1 shows the data points.

The following indices were used to evaluate the models' ability to predict STR in the presence of vegetation cover:

Table 8.1 Details of data points of ore grade

Parameter	Minimum	Average	Maximum	Standard deviation
Training	2	38	67	23
Testing	5	41	65	17

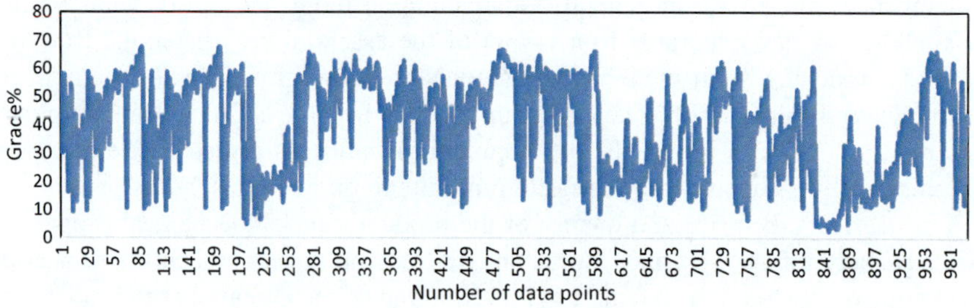

Fig. 8.1 The data points for iron ore grade

1. Root mean square error

$$\mathrm{RMSE} = \sqrt{\frac{\sum_{i=1}^{N} (\mathrm{OG_{me}} - \mathrm{OG_{es}})^2}{N}} \quad (8.1)$$

2. Mean absolute error (MAE)

$$\mathrm{MAE} = \frac{1}{N} \sum_{i=1}^{N} |\mathrm{OG_{me}} - \mathrm{OG_{es}}| \quad (8.2)$$

3. Percentage of bias (PBIAS)

$$\mathrm{PBIAS} = \frac{\sum_{i=1}^{N} (\mathrm{OG_{es}} - \mathrm{OG_{me}})}{\sum_{i=1}^{N} \mathrm{OG_{me}}} \quad (8.3)$$

4. Nash–Sutcliffe efficiency

$$\mathrm{NSE} = 1 - \frac{\sum_{i=1}^{N} (\mathrm{OG_{es}} - \mathrm{OG_{me}})^2}{\sum_{i=1}^{N} \left(\mathrm{OG_{me}} - \overline{\mathrm{OG}}_{\mathrm{me}}\right)^2} \quad (8.4)$$

5. Centered root mean square difference error (CRMSD)

$$\mathrm{CRMSD} = \sqrt{\frac{\sum_{i=1}^{N} \left(\mathrm{OG_{me}} - \overline{\mathrm{OG}}_{\mathrm{me}}\right) - \left(\mathrm{OG_{es}} - \overline{\mathrm{OG}}_{\mathrm{es}}\right)}{N}} \quad (8.5)$$

where $\mathrm{OG_{me}}$: Measured ore grade, $\mathrm{OG_{es}}$: estimated ore grade, N: number of data, $\overline{\mathrm{OG}}_{\mathrm{me}}$: average measured ore grade, N: number of data, and $\overline{\mathrm{OG}}_{\mathrm{es}}$: average estimated ore grade. The best model has the lowest RMSE, MAE, PBIAS, and CRMSE values. An ideal model has the highest NSE values.

8.2.2 Results

8.2.2.1 Selection of Random Parameters of Algorithms

This section aims to determine the best values for random parameters using a sensitivity analysis. The sensitivity analysis for population size (POP) is shown in

Fig. 8.2a. For the SSA, objective function values (RMSE) of the POP = 100, the POP = 200, the POP = 300, and the POP = 400 were 6.23, 5.12, 6.44, 6.58, and 6.98. Therefore, POP = 200 provided the lowest objective function for SSA. For the SCA, objective functions (RMSE) of the POP = 100, the POP = 200, the POP = 300, and the POP = 400 were 7.12, 7.01, 6.98, 7.23, and 7.34. Therefore, POP = 300 provided the lowest objective function for SSA. The sensitivity analysis for the maximum number of iterations (MNI) is shown in Fig. 8.2b. For the SASA, objective function values (RMSE) of the MNI = 80, the MNI = 160, the MNI = 240, and the MNI = 320 were 6.33, 5.16, 6.48, 6.59, and 7.01. Therefore, MNI = 160 provided the lowest objective function for SSA. For the SCA, the RMSEs of the MNI = 80, the MNI = 160, the MNI = 240, and the MNI = 320 were 7.19, 7.06, 6.95, 7.29, and 7.34.

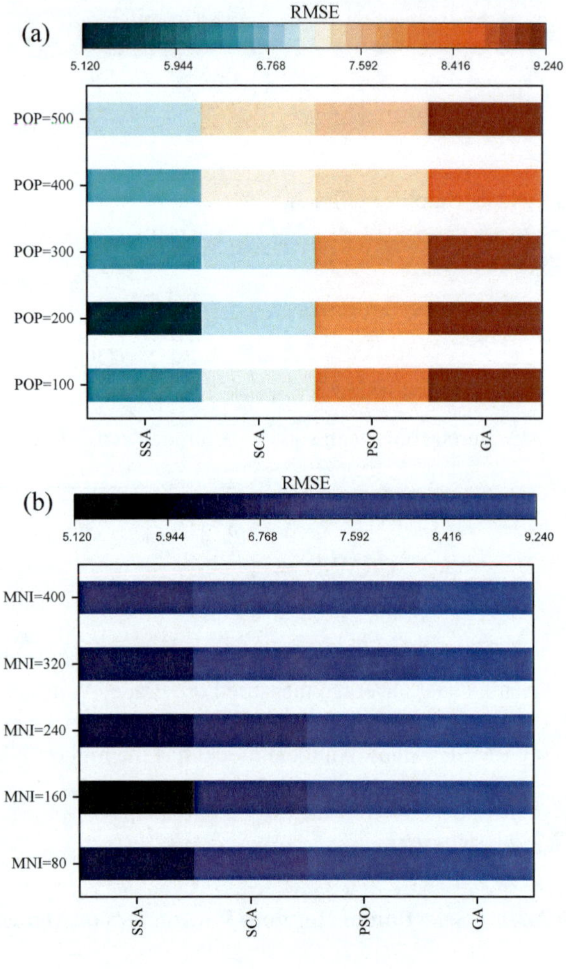

Fig. 8.2 Sensitivity analysis for **a** population size and **b** maximum number of iterations

8.2.2.2 The Investigation of the Accuracy of Models

We investigated the efficiency of models in this section. At the training level, the MAEs of the GMDH-SSA, GMDH-SCA, GMDH-FFA, GMDH-PSO, GMDH-GA, and GMDH were 4.55, 5.12, 5.54, 5.89, and 5.91. At the testing level, the GMDH-SSA decreased the MAEs of the GMDH-SCA, GMDH-FFA, GMDH-PSO, GMDH-GA, and GMDH by 4.7, 14, 16, and 17%, respectively. At the training level, the NSEs of the GMDH-SSA, GMDH-SCA, GMDH-FFA, GMDH-PSO, GMDH-GA, and GMDH were 0.95, 0.92, 0.88, 0.87, and 0.84. At the testing level, the NSEs of the GMDH-SSA, GMDH-SCA, GMDH-FFA, GMDH-PSO, GMDH-GA, and GMDH were 0.92, 0.90, 0.86, 0.82, and 0.80. At the training level, the PBIASs of the GMDH-SSA, GMDH-SCA, GMDH-FFA, GMDH-PSO, GMDH-GA, and GMDH were 5, 7, 12, 17, and 22. At the testing level, the PBIASs of the GMDH-SSA, GMDH-SCA, GMDH-FFA, GMDH-PSO, GMDH-GA, and GMDH were 7, 9, 14, 18, and 24 (Figs. 8.3 and 8.4).

Figure 8.4 shows boxplots of models for predicting iron ore grade. The medians of measured data GMDH-SSA, GMDH-SCA, GMDH-FFA, GMDH-GA, and GMDH models were 37.59, 37.05, 38.18, 39.81, and 41.54, respectively. Thus, the GMDH-SSA provided the highest accuracy among other models.

Figure 8.5 shows the CPU time of models. The CPU time of the GMH-SSA, GMDH-SCA, GMDH-PSO, GMDH-GA, and GMDH were 145, 198, 220, 232, and 242 s respectively.

8.3 Part B

8.3.1 Generalized Likelihood Uncertainty Estimation

A generalized likelihood uncertainty estimation (GLUE) estimates the uncertainty of a model prediction. Due to its flexibility and ease of implementation, the GLUE method has been widely used in various fields. Maghrebi and Vatanchi (2021) used GLUE to compute the parameter uncertainty of the rating curve. During the modeling process, model parameters provided high levels of uncertainty. Ehteram et al. (2022) used the GLUE model to estimate the uncertainty of model parameters and inputs of the ANN models. The input uncertainty was lower than the parameter uncertainty. Edwards et al. (2022) used GLUE to quantify the erosion model's uncertainty. Their results indicated that the models' uncertainty could be accurately reflected by GLUE. In this study, GLUE is used to quantify the uncertainty of GLUE parameters as follows:

(1) In the first step, the prior distribution of model parameters should be computed. The prior distribution of model parameters does not follow a normal distribution since model parameters lack a physical conception. The properties of MDH parameters can be determined by analyzing the variability of the parameters

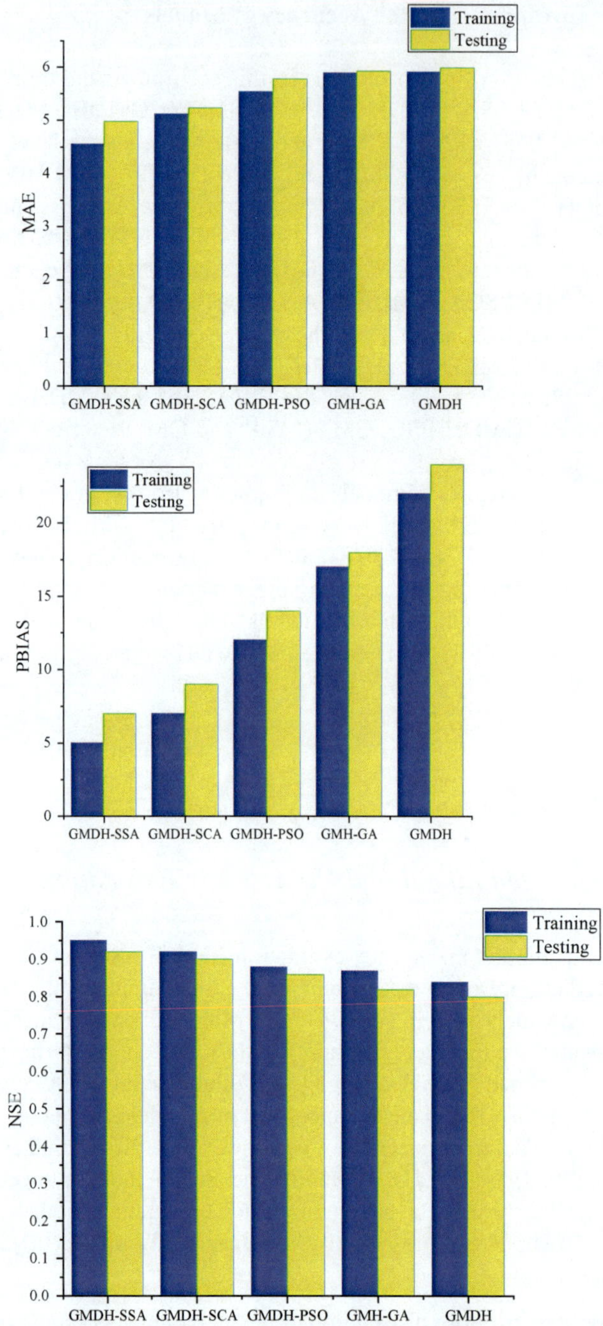

Fig. 8.3 The MAE, NSE, and PBIAS of the models for predicting iron ore grade

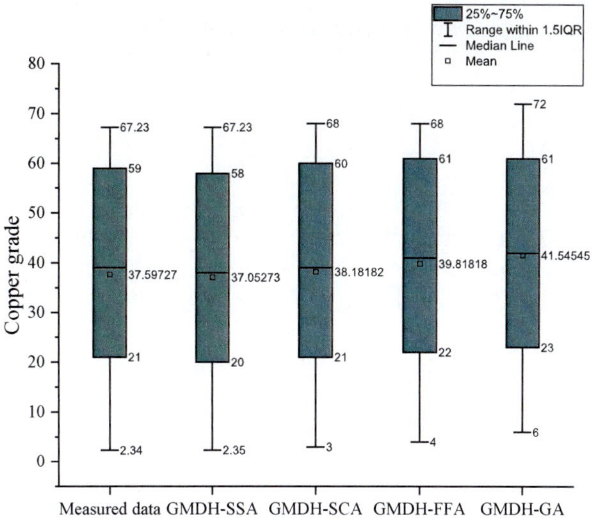

Fig. 8.4 The boxplots of models for estimating ore grade

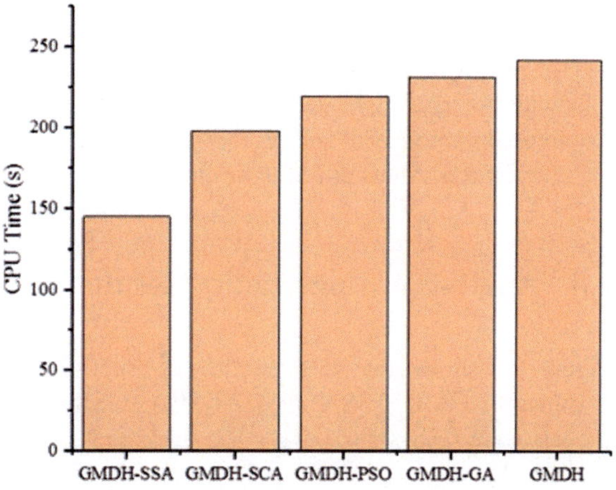

Fig. 8.5 The CPU time of models

during training. Hence, we used the trained 3000 GMDH models. The type of probability distribution and parameter values both remained unchanged between 1000 and 3000 GMDHs, so this number is sufficient.

(2) Monte Carlo provides many samples based on the prior parameter distributions.
(3) Create a model with randomly generated parameters (second step), and calculate the likelihood using the results (model errors).
(4) There are several types of likelihood functions. For example, the NSE can be used as a likelihood function.
(5) If we remove parameter sets with likelihood values below the threshold, we can get acceptable parameter sets as behavioral parameter sets.

(6) The posterior probability density is computed as follows:

$$p(\theta_i|\Upsilon) = \frac{p(\theta_i) * p(\Upsilon|\theta_i)}{\sum_{i=1}^{N} p(\Upsilon|\theta_i)} \tag{8.6}$$

where $p(\theta_i|\Upsilon)$: posterior probability density, $p(\Upsilon|\theta_i)$: likelihood values, and $p(\theta_i)$: prior parameter distributions.

(7) In this step, the parameters are estimated, and the means and variances are calculated.

(8) Two indices were used to quantify the uncertainty of models:

$$\mathrm{RB} = \frac{\sum_{t=1}^{n} (\mathrm{OG_U} - \mathrm{OG_L})}{N\sigma_o} \tag{8.7}$$

$$P = \frac{\mathrm{NO}}{N} \times 100 \tag{8.8}$$

where $\mathrm{OG_U}$ and $\mathrm{OG_L}$: upper and lower bounds of ore grade, NO: the number of observations enveloped by 95% prediction uncertainty (95PPU) band, RB: width of bound, *P*: percentage of the observed data that falls within the 95% prediction interval, and *N*: number of data. Models with the highest *P* and the lowest RB values have the best performances.

8.4 Analysis of Uncertainty Results

Figure 8.6 shows the 95% confidence interval. As observed in this figure, the P-values of GMDH-SSA, GMDH-SCA, GMDH-PSO, GMDH-GA, and GMDH models were 0.98, 0.96, 0.94, 0.92, 0.91, and 0.90, respectively. This figure shows that the RB values of GMDH-SSA, GMDH-SCA, GMDH-PSO, GMDH-GA, and GMDH models were 0.12 0.9615, 0.19, 0.21, 0.23, and 0.25, respectively. Thus, the GMDH-SSA and GMDH provided the lowest and highest uncertainties among other models.

8.5 Discussion on the GMDH Models

GMDH models were used to predict iron grade in the previous section. An optimization algorithm was used to find the GMDH model parameters accurately. Based on the results of the previous section, optimization algorithms play a significant role

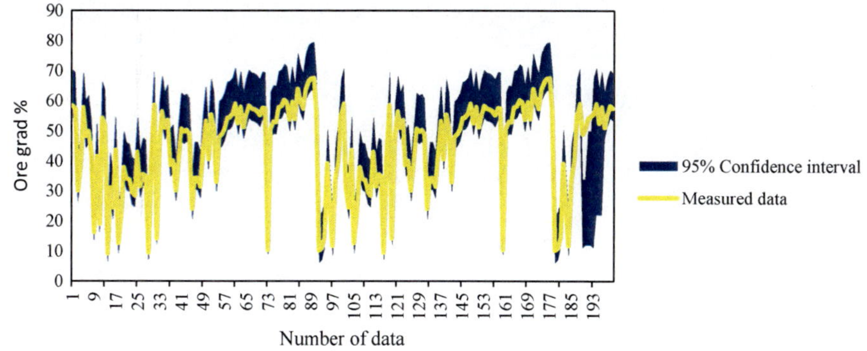

GMDH-SSA (P:0.98 and RB:0.12)

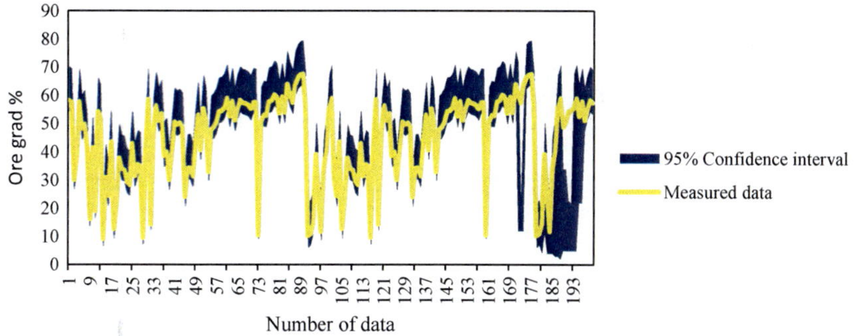

GMDH-SSA (P:0.96 and RB:0.15)

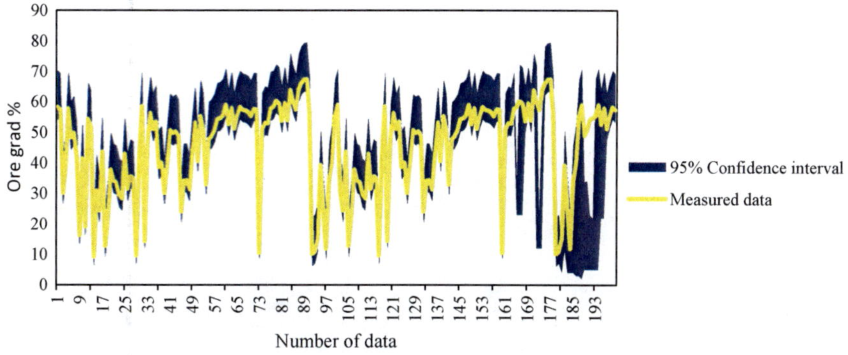

GMDH-SCA (P:0.94 and RB:0.19)

Fig. 8.6 95% confidence interval for the different models

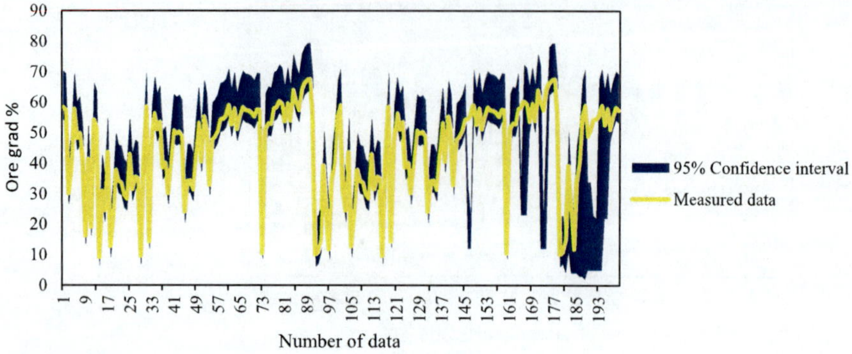

GMDH-PSO (P:0.92 and RB:0.21)

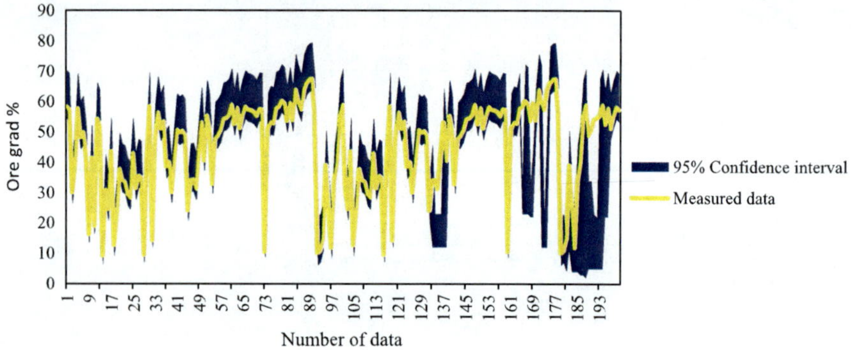

GMDH-GA (P:0.91, RBF:0.23)

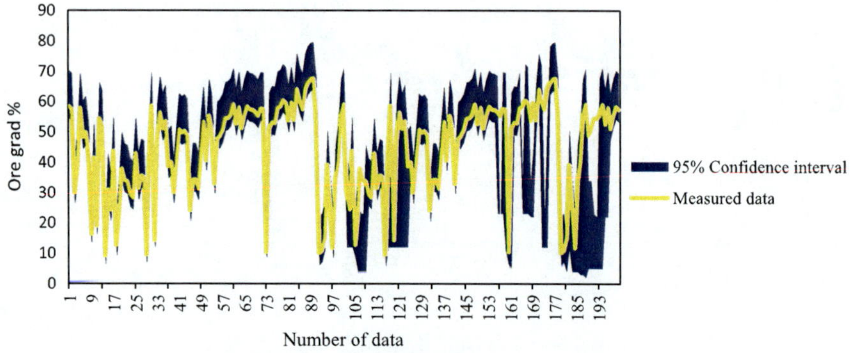

GMDH (P:0.90, RBF:0.25)

Fig. 8.6 (continued)

in improving the GMDH model's performance. The optimization algorithms utilize different operators, which results in different results. One of the most important tasks of modelers is using a robust training method. The models were trained using geographical coordinates. The GMDH models can accurately predict the iron grade without relying on geological data. The models can also be used for other purposes. GMDH hybrid models also required less computation time. By applying optimization algorithms, accuracy and timing are simultaneously improved. The optimization algorithms reduce the uncertainty of the GMDH model parameter estimation process by identifying the exact value of the GMDH model parameters.

8.6 General Remarks

GMDH models were used to predict the iron ore grade in this section. Results showed that the SSA method outperformed other optimization algorithms. A leader in the SSA structure directs other solutions toward the optimal ones. Results showed that improved GMDH models are very accurate in predicting iron grade. Geological maps can be prepared using the models described in this section. Iron-rich regions can be identified using these maps. Therefore, the models introduced in the present section can be used in the industry section.

References

Afeni, T. B., Lawal, A. I., & Adeyemi, R. A. (2020). Re-examination of Itakpe iron ore deposit for reserve estimation using geostatistics and artificial neural network techniques. *Arabian Journal of Geosciences*. https://doi.org/10.1007/s12517-020-05644-9

Bargawa, W. S., & Tobing, R. F. (2020). Iron ore resource modeling and estimation using geostatistics. *AIP Conference Proceedings*. https://doi.org/10.1063/5.0006928

Edwards, B. L., Webb, N. P., Galloza, M. S., Van Zee, J. W., Courtright, E. M., Cooper, B. F., Van Pelt, R. S., et al. (2022). Parameterizing an aeolian erosion model for rangelands. *Aeolian Research, 54*, 100769.

Ehteram, M., Panahi, F., Ahmed, A. N., Huang, Y. F., Kumar, P., & Elshafie, A. (2022). Predicting evaporation with optimized artificial neural network using multi-objective salp swarm algorithm. *Environmental Science and Pollution Research, 29*(7), 10675–10701.

Maghrebi, M. F., & Vatanchi, S. M. (2021). Uncertainty analysis of stage-discharge curves by generalized likelihood uncertainty estimation (GLUE) method. *Environmental Modeling & Assessment, 26*(4), 447–458.

Shahbeik, S., Afzal, P., Moarefvand, P., & Qumarsy, M. (2014). Comparison between ordinary kriging (OK) and inverse distance weighted (IDW) based on estimation error. Case study: Dardevey iron ore deposit, NE Iran. *Arabian Journal of Geosciences*. https://doi.org/10.1007/s12517-013-0978-2

Shamsi, R., Dehghani, H., Jalali, M., & Shokri, B. J. (2021). Ore grade estimation using the imperialist competitive algorithm (ICA). *Arabian Journal of Geosciences, 14*(14), 1–17.

Chapter 9
Conclusion and General Remarks for Estimating Ore Grade

Abstract This chapter investigates the capabilities of different models for estimating ore grades. The features of different methods are explained. The advantages and disadvantages of different models are described. This chapter presents the solutions for improving the accuracy of soft computing models. This chapter explains the details for quantifying uncertainty modeling. The chapter indicated that the artificial neural network models (ANN) had high capabilities for estimating ore grades. This chapter indicated that the model parameters and input parameters are the uncertainty resources in the modeling process. Also, the optimization algorithms improved the accuracy of ANN models for estimating ore grade.

Keywords Optimization algorithms · Ore grade estimation · ANN models · Uncertainty resource

9.1 Introduction

The growth of industrial units, population, and economic development increases the demand for raw materials. Mineral potential assessment is an important step in mining engineering. The exploration and exploitation of a mine are expensive, so identifying potential mine areas plays a key role in reducing exploration costs (Liu et al., 2021). Grade estimation is a fundamental aspect of mining engineering. Different models can be used to estimate ore grade. This process is complex and nonlinear. Grade estimation is influenced by various factors, such as the geological structure of the region, geological processes, and various interactions (Shamsi et al., 2021). Finding the relationship between ore grade and various input parameters is difficult and complex. Classical models such as regression and kriging can estimate grade, but they aren't accurate enough. As a result, mining engineers must develop previous models and introduce new models to estimate grade (Abbaszadeh et al., 2021). This process may pose numerous challenges.

Identifying key input parameters is an important issue in the modeling process. The next step is to prepare and select a suitable model for estimating the grade. It is challenging to estimate grades accurately. Mining engineers and modelers can solve these challenges by selecting appropriate models and identifying appropriate inputs.

M. Ehteram et al., *Estimating Ore Grade Using Evolutionary Machine Learning Models*, https://doi.org/10.1007/978-981-19-8106-7_9

Artificial intelligence methods play an important role in technology advancement. Artificial intelligence methods' high accuracy and speed allow modelers to produce accurate results. Therefore, these models play a significant role in developing mining engineering. The artificial neural network is a well-known soft computing model. This model consists of small units called neurons. Throughout the neural network, neurons serve as the processing units. Models like kriging are useful for linear analysis, but soft computing models are useful for linear and nonlinear analysis. There are several types of artificial neural network models. They usually use training and testing data to estimate outputs. Determining model parameters is one of the most important challenges of ANN models. Choosing the best input scenarios is another challenge for artificial neural network models. Model parameters must be correctly set to maximize the accuracy of artificial neural network models. A properly adjusted neural network model can reduce the uncertainty of its parameters. Using optimization algorithms can help improve the accuracy of artificial neural networks. Optimization algorithms can improve the accuracy of AI models. Algorithms can connect with soft computing models easily.

9.1.1 Final Results

In this study, optimized multi-layer perceptron (MLP), radial basis function neural network (RBFNN), extreme learning machine model (ELM), and group machine handling data (GMDH) were used to estimate grade. The following points should be taken into account based on the results of the previous sections:

1. Although optimization algorithms have a high capacity for finding optimal values of model parameters, their results are not definite and contain some uncertainties. Therefore, one of the sources of uncertainty is the parameters of the models.
2. Previous research mainly ignored the sources of uncertainty in modeling. The current study examined the sources of uncertainty. Optimized models had less uncertainty than other models.
3. Increasing the number of model parameters may increase uncertainty.
4. Selecting optimization algorithms to train models is one of the most important issues in the modeling process.
5. Neural network models can improve the accuracy of classical models such as the Kriging model. Hybridizing neural networks and kriging models improve the accuracy of both models.
6. Using ensemble models can increase the accuracy of individual models.
7. Bayesian model, as one of the ensemble models, can increase the accuracy of individual models.

8. Inclusive multiple models (IMM) can be used as an ensemble model to estimate the grade. The IMM model has no computational complexity.
9. Using geographical parameters as input scenario produces accurate results.
10. The use of optimization algorithms reduces the computational time of models.

9.1.2 *Suggestions for the Future Studies*

The present study could be developed in future research:

1. Different tests can be used to select the best input scenario.
2. Other soft computation models can also be used to estimate grades.
3. The models of the current book should be tested for other case studies worldwide.
4. Using neighborhood points will improve the accuracy of the models.
5. The effect of Input parameter uncertainty on output should be assessed.
6. In this study, optimization algorithms were used to find the weight connection of the neural network. Some weight connections have small amounts, which only increases the computational time of the neural network. Reducing computational time requires identifying redundant connection weights.

References

Abbaszadeh, M., Ehteram, M., Ahmed, A. N., Singh, V. P., & Elshafie, A. (2021). The copper grade estimation of porphyry deposits using machine learning algorithms and Henry gas solubility optimization. *Earth Science Informatics*. https://doi.org/10.1007/s12145-021-00667-6

Liu, Z. N., Yu, X. Y., Jia, L. F., Wang, Y. S., Song, Y. C., & Meng, H. D. (2021). The influence of distance weight on the inverse distance weighted method for ore-grade estimation. *Scientific Reports*. https://doi.org/10.1038/s41598-021-82227-y

Shamsi, R., Dehghani, H., Jalali, M., & Shokri, B. J. (2021). Ore grade estimation using the imperialist competitive algorithm (ICA). *Arabian Journal of Geosciences*. https://doi.org/10.1007/s12517-021-07808-7